Biotech's

DICTIONARY *Of* Horticulture

Compiled & Edited by

Dinesh Arora

BIOTECH

2011

BIOTECH BOOKS

DELHI - 110035

Biotech's
DICTIONARY
Of
Horticulture

Compiled & Edited by
Dinesh Arora

ISBN 81-7622-119-8

Published by : **BIOTECH BOOKS**
1123/74, Tri Nagar,
Delhi - 110 035
Phone: 27382765
e-mail: biotechbooks@yahoo.co.in

Showroom : 4762-63/23, Ansari Road, Darya Ganj,
New Delhi - 110 002
Phone: 23245578, 23244987

Printed at : **Chawla Offset Printers**
New Delhi - 110 052

PRINTED IN INDIA

PREFACE

The word Horticulture comes from two Latin words, hortus, 'garden' and cultura, 'cultivation'. Horticulture is the science and art of growing fruits, vegetables, flowers, shrubs and trees. Originally, Horticulture meant the practice of gardening and extending from the same, it now means the cultivation of plants that were once grown in gardens. People generally get confused between the terms agriculture and horticulture. The term agriculture refers to more open forms of culture like the production of grains and grasses that are cultivated on a large scale. Nonetheless, the original distinction between the two have become so blurred that many crops that were initially considered either agronomic or horticultural, are now sometimes taken in one field and sometimes in the other, as per their intended use. Therefore, a plant grown for home consumption may be called horticultural, while the same plant cultivated for forage is regarded as an agronomic crop.

Horticulture consists of several branches, viz. Pomology, which means the growing of fruit (especially tree fruits); Olericulture, i.e. the production of vegetable crops; Floriculture, which is the production of flowers; and Ornamental Horticulture, also known as landscape gardening, which includes the maintenance and design of home grounds, public gardens and parks, private estates, botanical gardens, and recreational areas such as golf courses, football fields, and baseball diamonds.

The intent behind making the dictionary is to provide the readers with a storehouse of information on the jargon used in the discipline. The use of simple and comprehensible language, along with pictures and examples, makes it a must buy for those who are venturing into the fields as well as those who are already a part of it.

- **a horizon**

the uppermost layers of soils consisting of partly decomposed plant remains and relatively fresh leaves and other plant debris. The surface mineral layer, high in organic matter and dark in colour, and the lighter coloured layer where leaching of solutes and suspended materials occur.

- **abaxial**

directed away from the stem of a plant, pertaining to the lower surface of a leaf.

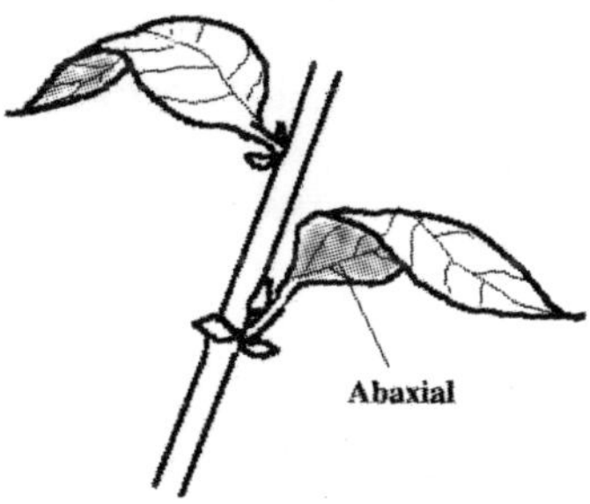

- **aberrcnt**

different from the normal type of species, genus, or higher group in one or more characters, but not readily assignable to another group.

- **abscission**

of plants: The shedding of leaves or other parts as the result of physical weakness in a specialised layer of cells (abscission layer that develops at the base).

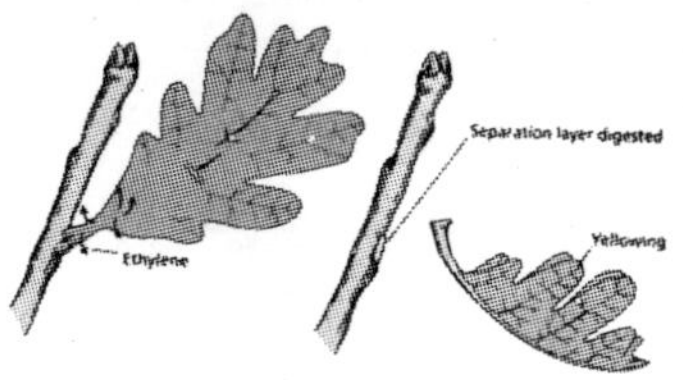

- **absidia**

a zygomycete fungus. Reported to be allergenic. May cause mucorosis in immune compromised individuals. The sites of infection are the lung, nasal sinus, brain, eye and skin. Infection may have multiple sites.

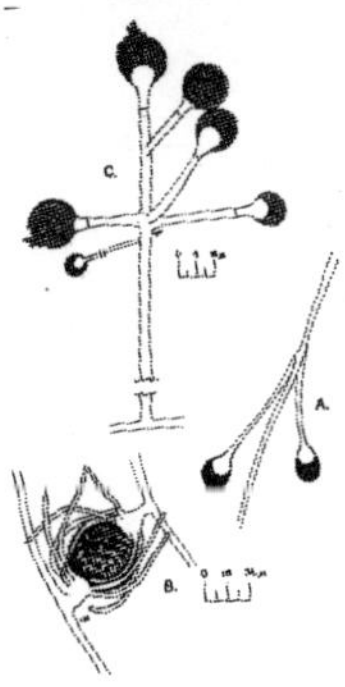

- **absorption**

movement of ions and water into an organism as a result of metabolic processes, frequently against an electrochemical potential gradient (active) or as a result of diffusion along an activity gradient (passive).

- **acaulescent**
  having no visible stem, or a very short one.

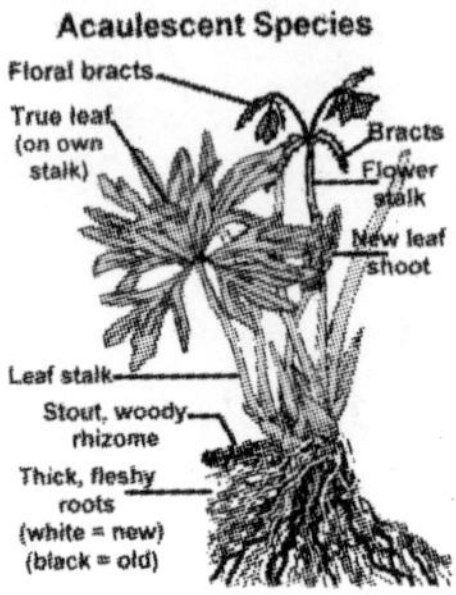

- **acaulis**
  having no stem.

- **accent plant**
  this could be a focal point plant. A plant to catch attention. Could also be called an anchor plant.

- **accrete**
  grown together.

- **acerose**
  needle-shaped, like pine needles.

- **acetylene-block assay**
  estimates denitrification by determining release of nitrous oxide (N_2O) from acetylene-treated soil.

- **acetylene-reduction assay**
  estimates nitrogenase activity by measuring the rate of acetylene reduced to ethylene.

- **acheve**
  a one seeded fruit which does not split open to release its seed, e.g., the 'seeds' on a strawberry.

- **achlamydeous**
  of flowers: lacking petals and sepals, e.g., willow (Salix spp.).

- **acicular**
  needle like spine, pointed, bristle.

- **acid**
  soil or compost with a pH level of less than 7.

• **acid medium**
a compost which contains little or no lime and has a pH of less than 6.5. Sometimes referred to as 'sour' soil by gardeners.

• **acid rain**
rainwater that contains sulphur dioxide and other pollutants from industrial plants. There has been considerable damage done to the forests in different parts of the world.

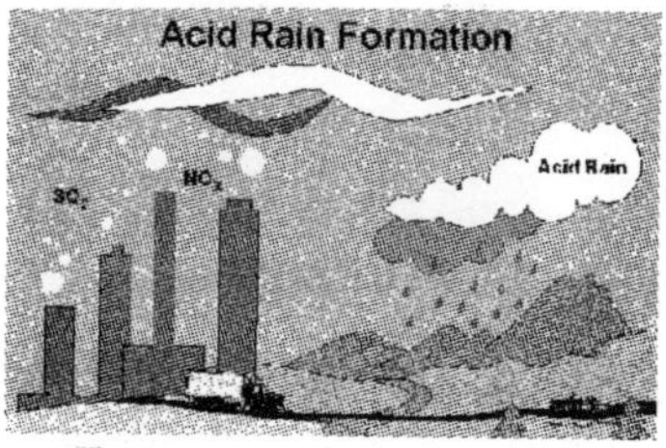

• **acid soil**
soil that is lower than 7.0 pH (higher would be alkaline). Acidity is measured by the amount of calcium in the soil, as is alkaline soil.

• **acid-detergent fibre**
insoluble residue following extraction with acid detergent (Van Soest method), i.e., cell wall constituents minus hemicelluloses.

• **acranthous**
term applies to sympodial type orchids, referring to the annual portions of successive growth of the rhizome, each beginning with scaled-leaves, ending with an inflorescence.

• **acropetal**
describes the development of structures (such as spores in succession from the base towards the apex).

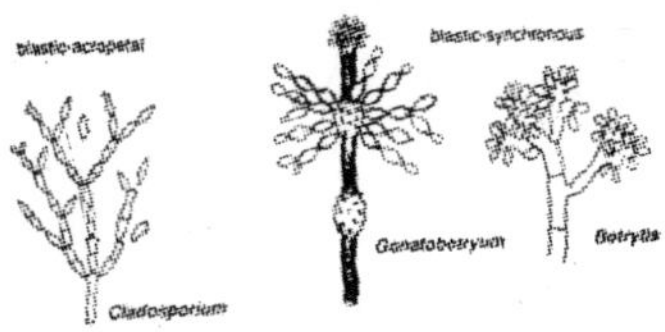

• **actinomycete**
non-taxonomic term applied to a group of high G + C base composition, Gram-positive bacteria that have a superficial resemblance to fungi. Includes many but not all organisms belonging to the order Actinomycetales.

• **active site**
region of an enzyme where substrates bind.

• **aculeate**
prickle-shaped.

• **acumen**
a tapering point.

- **adaxial**
directed toward the stem of a plant, pertaining to the upper surface of a leaf.

- **Adenosine triphosphate (ATP)**
common energy-donating molecule in biochemical reactions. Also an important compound in transfer of phosphate groups.

- **adjuvant**
material added to improve some chemical or physical property (e.g., of a plant) protectant or a biological property (e.g., to improve antibody response to an antigen).

- **adnate**
mycology of gills or tubes broadly attached to the stipe, attached by nearly the entire width of the gills or layer of tubes.

- **adnation**
adnate fusion of unlike parts, e.g. labellum with column, contrasted with connation.

- **adsorption**
process by which atoms, molecules, or ions are taken up and retained on the surfaces of solids by chemical or physical binding.

- **adumbral**
shady.

- **adventitious**
plant parts, such as shoots and roots, produced in an unusual position on a plant or at an unusual time of development.

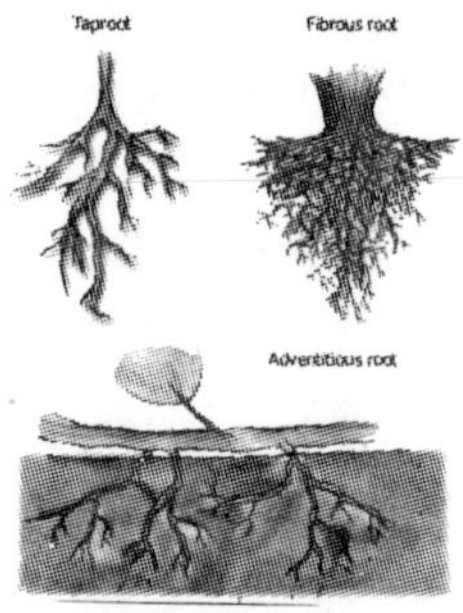

- **adventitious plant**
a young plant that develops in an asexual manner on the leaves or stems of the mother plant.

- **aecidium**
an aecium with a cup-like outer wall.

- **aeciospore**
a dikaryotic 'transfer' spore of the Uredinales, formed in an aecium's on the alternate host in macrocyclic rusts and infecting only the primary host. Spores of

Stage I in heteroecious or autoecious rusts.

- **aecium**

in the Uredinales the first sorus that is formed after plasmogamy and bears binucleate aeciospores (Stage I) heteroecious or autoecious rusts.

- **aeration**

loosening compacted soil to allow air to circulate. It also applies to lawns when a fork or other implement is used to svab holes in it, also to allow air in.

- **aerial root**

a root which grows out from the stem above ground level. Aerial roots are commonly seen on mature specimens of Monstera deliciosa.

- **aerobe**

an organism that requires free oxygen for respiration

- **aerobic**

1. having molecular oxygen as a part of the environment.
2. growing only in the presence of molecular oxygen, as in aerobic organisms.
3. occurring only in the presence of molecular oxygen. as in certain chemical or biochemical processes such as aerobic respiration.

- **aerotolerant anaerobes**

microbes that grow under both aerobic and anaerobic conditions, but do not shift from one mode of metabolism to another as conditions change. They obtain energy exclusively by fermentation.

- **aflatoxin**

a polynuclear substance and a known carcinogen derived from moulds, specifically, that is produced by a fungus occurring on peanut.

- **agamospermy**

all types of apomixes in which seeds are formed by asexual means. Does not include vegetative reproduction.

- **agar**
complex polysaccharide derived from certain marine algae that is a gelling agent for solid or semisolid microbiological media. Agar consists of about 70% agarose and 30% agaropectin. Agar can be melted at temperature above 100 °C, gelling temperature is 40-50° C.

- **agar-agar**
gelantinous substance obtained mostly as transluscent strips or white powder from certain sea weeds, used as solidifying agent in culture media.

- **agarose**
one of the constituents of agar. Often used in preference to agar because it gels at a lower temperature and does not contain the inhibitors of virus growth frequently present in agar. It is also used widely in gel electrophoresis because it has a more uniform pore size than that of agar.

- **agent of disease**
an organism or abiotic factor that causes disease, a pathogen.

- **agent of innoculation**
that which transports innoculum from its source to or into the infection court (e.g., wind, splashing rain, insects, humans)

- **agglutination**
the formation of insoluble aggregates following the combination of antibodies with cells or other particulate antigens or with soluble antigens bound to cells or other particles or following the combination of soluble (or particulate) antigens with cell-bound or particle-bound antibodies.

- **aggregate culture**
the use of solid material to grow plants. Some examples are: gravel, rock wool, sand, all with the additional use of a nutrient-water soluble solution.

- **aggregates**
clumps or cemented units of mineral and organic matter.

- **aggressiveness**
of a plant pathogen: relative ability to colonise and cause damage to plants.

- **agro-forestry**
land-use system in which woody perennials are grown for wood production with agricultural crops, with or without animal production.

- **air layering**
a method of propagating single-stem plants, such as *ficus elastica decora*, which have lost their lower leaves and become leggy. An incision is made to a portion of outer stem layer, damp sphagnum moss is wrapped in a bag around it until roots develop. Then it is cut and replanted with its shorter stem size.

- **akinete**
thick-walled resting cell of cyanobacteria and algae.

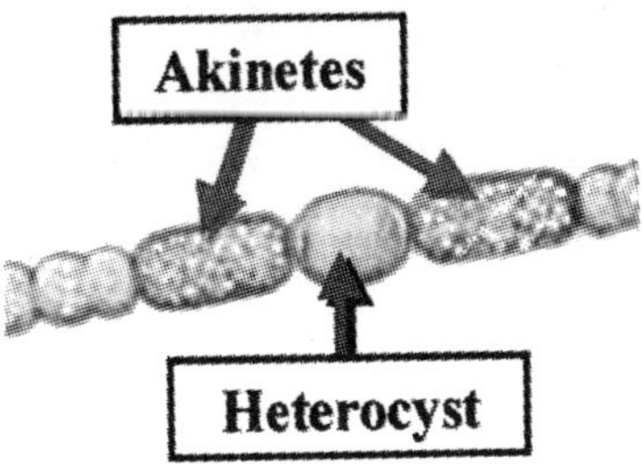

- **alae**
expansions or projections formed by a longitudinal thickening of the cuticle of a nematode. Cervical alae are confined to the anterior region of nematodes parasitic in animals. Caudal alae occur in the posterior region of males in a number of genera. Longitudinal alae, usually four, extend the length of the body sub laterally.

- **alata**
winged.

- **alba**
flower with all segments white, but which may have some degree of yellow on the lip only.

- **albescent**
becoming white or yellow.

- **albinism**
lack of colour. Deficient in pigment.

- **aleurone**
a layer of protein granules found in the endosperm of ripe seeds or fruits.

- **alga**
phototropic eukaryotic micro organism. Algae could be unicellular or multicellular. Blue-green algae are not true algae, they belong to a group of bacteria called cyanobacteria. (plural, algae)

- **aliphatic**
organic compound in which the main carbon structure is a straight chain.

- **alkaline**
soil that has a pH level of about 7 or more.

- **alkaline soil**
soil that has a pH level of about 7.0 or more. Sometimes referred to as 'sweet' soil by gardeners.

- **alkalophile**
organism that grows best under alkaline conditions (up to a pH of 10.5).

- **alkane**
straight chain or branched organic structure that lacks double bonds.

- **alkene**
straight chain or branched organic structure that contains at least one double bond.

- **allantoid**
of spores: sausage-shaped, somewhat curved, with rounded ends.

- **allee**
1. a very formal design of planting trees lining both sides of a path or drive.
2. any of one or more alternative forms of a given gene, both (or all) alleles of a given gene are concerned with the same trait or characteristic, but a particular allele codes for a product qualitatively and/or quantitatively different from that coded by other alleles of that gene.

- **allelomorph**
see **allele**.

- **allelopathy**
the release of chemicals by certain plants that will prevent the growing of other plants nearby. Walnut trees are very well known to do this.

- **alliance**
designates a group of genera that have many common characteristics and can be used for cross breeding to produce new hybrid gen-

era. An alliance is limited to genera within a single tribe.

- **allo**
  combining form denoting differential characteristics or forms, differentiation from normal

- **allochthonous flora**
  organisms that are not indigenous to the soil but that enter soil by precipitation, diseased tissues, manure, and sewage. They may persist for some time but do not contribute in a significant way to ecologically significant transformations or interactions.

- **alloplasm**
  cytoplasm from an alien species that has been transferred by backcrossing into a cultivated species. The alien cytoplasm's most common and readily observable effects are male sterility, female sterility, or reduced plant vigour. If the alien cytoplasm does not manifest itself, the cytoplasm is not considered alien.

- **alpine**
  plants from high mountain regions. Anything that is from above the tree line. They are able to survive winter beneath deep snow protected from extreme low temperature by their moisture.

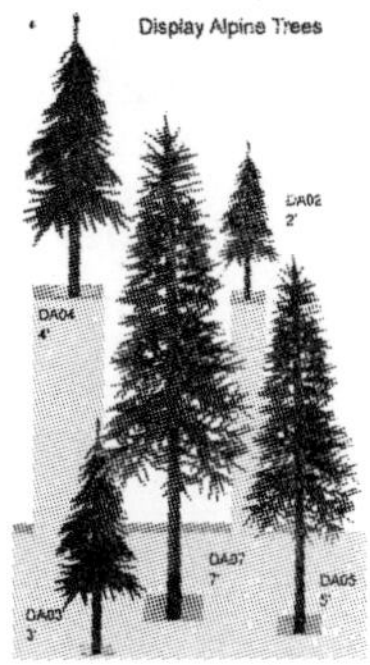

- **alpine house**
  a special greenhouse created to meet the requirements of alpines. It is usually kept cool in the summer by shading the glass. In the winter it is unheated unless to provide protection from very severe cold.

- **alternate**
  leaf form, where the leaves are arranged singly at different heights on the stem.

- **alternative host**
  a plant other than the main host that a parasite can colonise, alternative hosts are not required for completion of the developmental cycle of the parasite

- **amendment, physical**
any substance (such as sand, calcined clay, peat, or sawdust) added to soil for the purpose of altering physical conditions.

- **amensalism (antagonism)**
production of a substance by one organism that is inhibitory to one or more other organisms. The terms antibiosis and allelopathy also describe cases of chemical inhibition.

- **amerospore**
a non-septate spore.

- **amerosporous**
having one-celled spores.

- **amino group**
an NH_2 group attached to a carbon skeleton as in the amines and amino acids.

- **ammonification**
liberation of ammonium (ammonia) from organic nitrogenous compounds by the action of micro-organisms.

- **amphibious**
the ability of plants to grow both in aquatic and in the exposed soil. Usually in a moist or boggy condition, when the winter recedes in the area.

- **amphid**
in nematodes: a chemosensory organ, occurring laterally in pairs, located in the anterior region. Sometimes called lateral organs.

- **amphidelphic**
in nematodes: having two ovaries, one directed anteriorly and the other posteriorly.

- **amphidiploid**
a plant possessing the sum of the somatic chromosome number of two species.

- **amphigynous**
in fungi: having an antheridium through which the oogonial initially grows.

- **amphiphloic**
having both internal and external phloem.

- **amphivasal bundle**
concentric vascular bundle in which the xylem surrounds the phloem, as in the rhizomes of some monocots.

- **amplexicaul**
clasping the stem.

- **amplification**
production of multiple copies of a DNA sequence, either in vivo or in vitro, starting with one or a few copies.

- **ampulliform**
flask-like in form.

- **amyloid**
of spore walls, spore ornamentation, hyphal walls, ascus tips, etc.: staining blue or greyish to blackish-violet in Melzer's reagent, presumably because of the presence of starch or a starch-like compound.

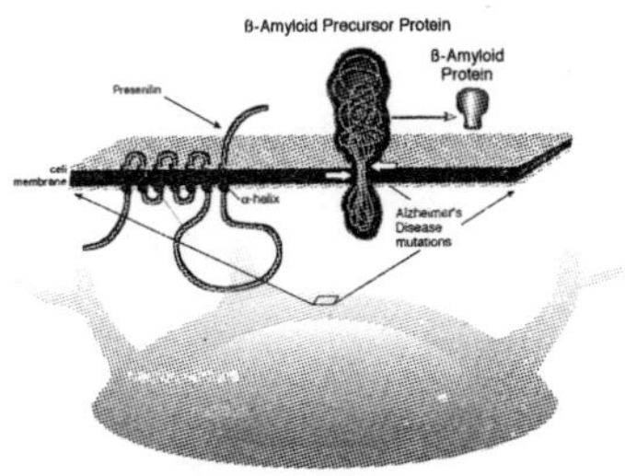

- **anabolism**
metabolic processes involved in the synthesis of cell constituents from simpler molecules. An anabolic process usually requires energy.

- **anaerobic**
bacteria, etc, that occur when there is no oxygen. It happens sometimes in stagnant ponds or in unhealthy compost heaps when they get cold and damp.

- **anaerobic respiration**
metabolic process whereby electrons are transferred from an organic, or in some cases, inorganic compounds to an inorganic acceptor molecule other than oxygen. The most common acceptors are nitrate, sulphate. and carbonate.

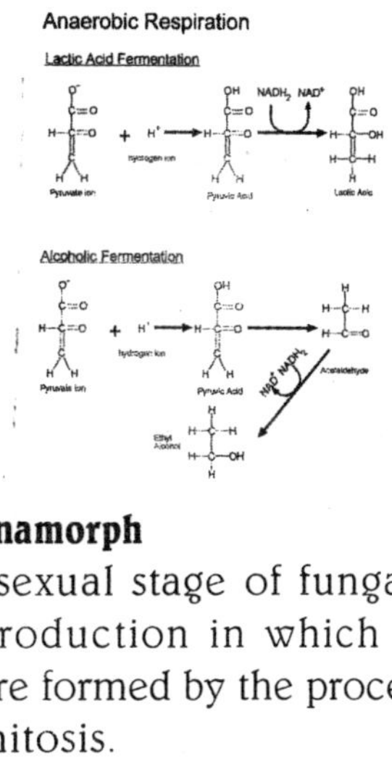

- **anamorph**
asexual stage of fungal reproduction in which cells are formed by the process of mitosis.

- **anastomosis**
fusion of somatic hyphae, characteristic of most dikaryomycota. (pl. anastomoses)

- **androdioecious**
  plants that have male and perfect flowers on separate plants.

- **andromonoecious**
  a plant bearing bisexual or perfect flowers in addition to staminate flowers.

- **aneuploid**
  chromosome constitution differing from the usual diploid constitution by loss or duplication of chromosomes or chromosomal segments.

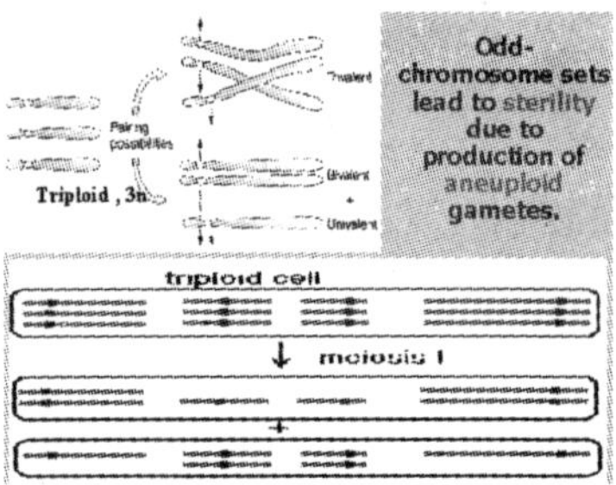

- **angiocarpous**
  of basidium-producing organs: hymenial surface at first exposed but later covered by an incurving pileus margin and/or excresences from the stipe.

- **angulate**
  more or less angular.

- **anion exchange capacity**
  sum total of exchangeable anions that a soil can adsorb. Expressed as centimoles of negative charge per kilogram of soil.

- **anisogamy**
  sexual reproduction in which the female gamete is large and immobile and the male gamete is small and motile.

- **annual**
  a plant whose entire life cycle, from germination to making seeds and dying, is completed in one season.

- **annule**
  thickened interval between transverse striae in the cuticle of a nematode.

- **annulus**
  a membraneous skirt surrounding the stipe of a hymenomycete or gasteromycetepl.

- **anoxic**
  literally 'without oxygen.' An adjective describing a microbial habitat devoid of oxygen.

- **anoxygenic photosynthesis**
  type of photosynthesis in green and purple bacteria in

which oxygen is not produced.

- **antagonism**
an ecological association between organisms in which one or more of the participants is harmed or has its activities limited.

- **antagonist**
biological agent that reduces the number or disease-producing activities of a pathogen.

- **antagonistic symbiosis**
parasitism, one organism of an association benefits at the expense of the other.

- **anther**
the part of the flower which produces pollen. It is the upper section of the stamen.

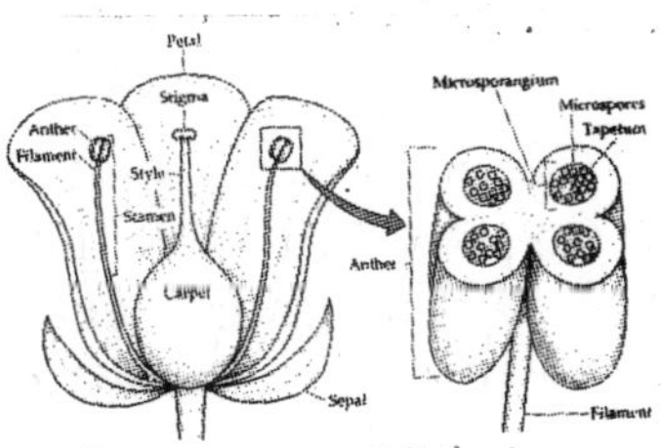

- **antheridium**
male gametangium found in the phylum Oomycota (Kingdom Stramenopila) and phylum Ascomycota (Kingdom Fungi).

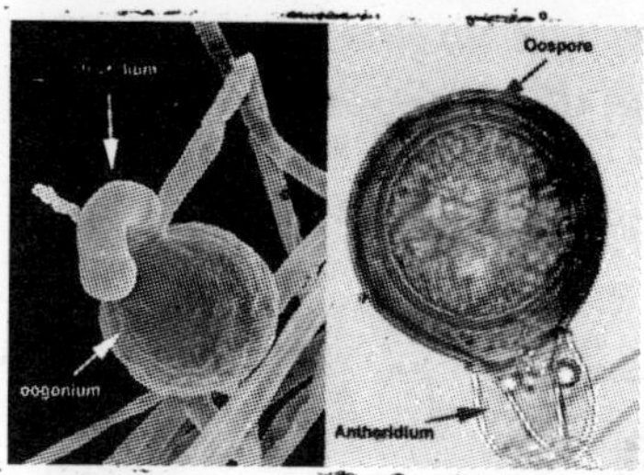

- **anthesis**
the action or period of opening of a flower, especially the period when the stigma is favourable for the germination of pollen coming in contact with it.

- **anthocyanescence**
having reddish-purple colour in tissues that are normally green, often a symptom of plant disease appearing in the nearly dead margins around completely dead spots in green leaves

- **anthoxanthins**
water-soluble pigments widely distributed in leaves, stems, and flowers. They are mostly colourless, but can modify other colours, especially anthocyanins.

- **anthracnose**
any of various plant diseases, particularly those caused by fungi of the

Melanconiales, in which discrete, dark-coloured, necrotic lesions develop on the leaves, stems, and/or fruits.

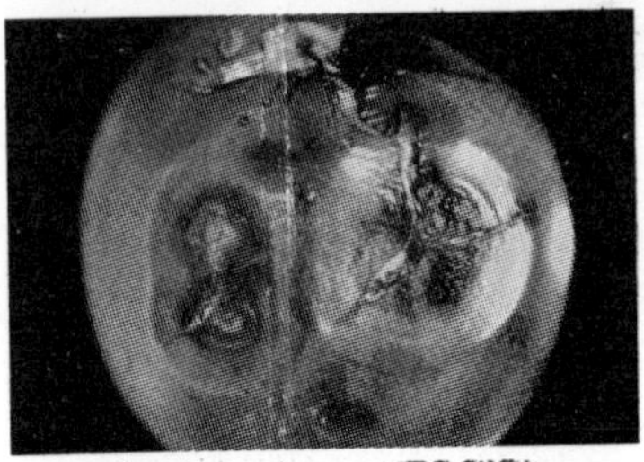

- **antibiotic**

organic substance produced by one species of organism that in low concentrations will kill or inhibit growth of certain other organisms.

- **antibody**

any immunoglobulin molecule produced in direct response to an antigen and which can combine specifically, non-covalently, and reversibly with the antigen which elicited its formation.

- **anticous**

the fore-part, i.e. that most remote or turned away from the axis.

- **antigen**

molecule of carbohydrate or protein which stimulates the production of an antibody, with which it reacts specifically.

- **antioxidant**

an organic compound that accepts free radicals and thus prevents autoxidation of fats and oils. At very low concentrations in food, antioxidants not only act in retarding rancidity but protect the nutritional value or minimise the breakdown of vitamins and essential fatty acids.

- **antiseptic**

agent that kills or inhibits microbial growth but is not harmful to human tissue.

- **antiserum**

the serum from a vertebrate that has been exposed to an antigen and which contains antibodies that react specifically with the antigen.

- **aphids**

small sap sucking insects. They infect foliage and are easily recognised by the sugary 'honey dew' that they secrete that often attracts ants. Can be controlled by application of soapy solution.

- **aphyllous**
  without leaves.
- **apical**
  at the tip, as in an inflorescence borne at the top of the stem or pseudobulb.

Apical inflorescence

- **apical bud**
  a bud at the apex or terminal position on a plant or branch.
- **apiculate**
  having an apical point or apiculus.
- **apiculus**
  a short, sharp, but not stiff, point, usually as the bud of a spore.
- **apoplast**
  the continuum of nonliving plant cell-wall material that surrounds the symplast. It includes the tracheary elements of xylem and the free space, and is one of two parallel pathways for solute movement through plants (the other being symplast). Substances can move through the apoplast only down a thermodynamic potential gradient.
- **apothecium**
  an ascus-bearing structure (ascocarp) in which the ascus-producing layer (hymenium) is not covered by fungal tissue at maturity.

Apothecium

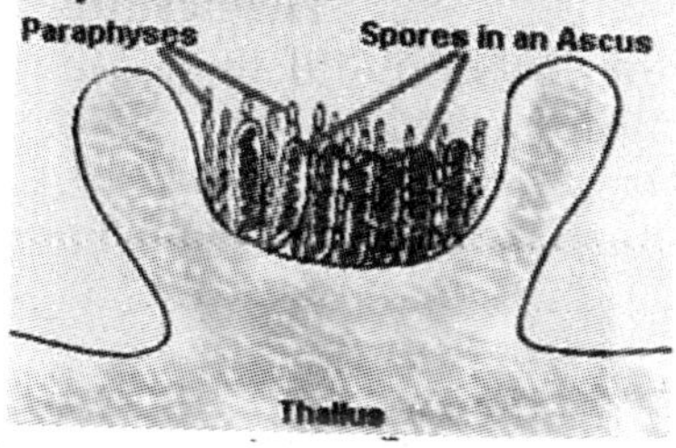

- **apparent free space**
  a measure of free space gained from ion influx and efflux experiments. It is influenced by ion exchange of solutes with cell walls and thus can differ for each solute.
- **applanate**
  flattened out or horizontally expanded.

- **appressed**
  lying flat for the whole length of the organ.

- **appressorium**
  an enlargement on a hypha or germ tube that attaches itself to the host before penetration takes place.

- **aquatic**
  plant which grows partially or completely in water.

- **arbor**
  a free standing structure used in the garden to support vines or climbing plants of all sorts for shade, a walkway, or just a focal point. This term is sometimes used interchangeably with pergola.

- **arboretum**
  this is a garden with a large collection of trees and shrubs. They are specifically grown for scientific or educational purpose.

- **arborist**
  a specialist who cares for and maintains trees. Everything from planting to pruning and also diagnosing and treating diseases is his responsibility.

- **arbuscular mycorrhiza**
  mycorrhizal type that forms highly branched arbuscules within root cortical cells.

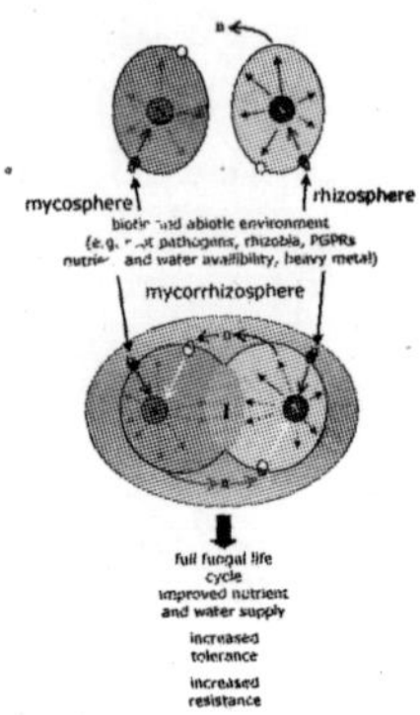

- **arbuscule**
  of vesicular-arbuscular mycorrhizae: a much-branched, microscopic haustorial structure of the fungal symbiont that forms within living cortical cells of the root. The interface of the arbuscule with the plant protoplast is a site of exchange of nutrients and growth-regulating chemicals

- **archaea**
  evolutionarily distinct group (domain) of prokaryotes consisting of the methanogens, most extreme halophiles and hyperthermophiles, and thermoplasma.

- **archaebacteria**
older term for the Archaea.

- **arcuate**
curved like a bow.

- **areolated**
divided into small spaces or areolations, usually pertains to the cuticle of a nematode.

- **areolation**
a condition in which the transverse body annulation of a nematode traverses the lateral field.

- **areole**
a small well-defined area, usually hairy and cushion-like, found on the stem of cacti. From them arise spines or glochids.

- **aristate**
tipped with bristle-like appendage or awn.

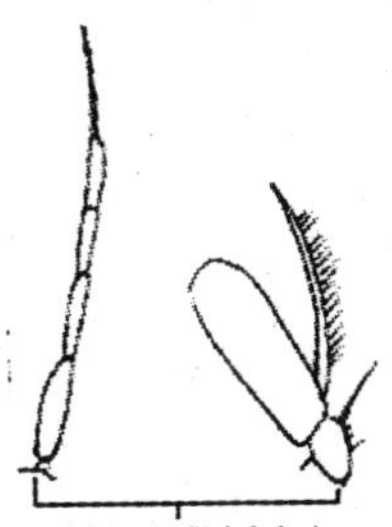

- **aroma**
a characteristic odour, as of a plant, feed, or food.

- **aromatic**
organic compounds which contain a benzene ring, or a ring with similar chemical characteristics.

- **arthropod**
invertebrate animals (insects, arachnids and crustaceans) that have a jointed body and limbs and usually a hard shell or exoskeleton that is molted periodically.

- **arthrospore**
a spore resulting from the fragmentation of a hypha, as in the conidial stage of many Basidiomycetes.

- **artificial turf**
a fabricated rug of fibres simulating turf. Usage: Astroturf is a trademark, not the general term.

- **ascocarp**
ascospore-bearing, multi cellular sporocarp formed by a member of the Ascomycotina.

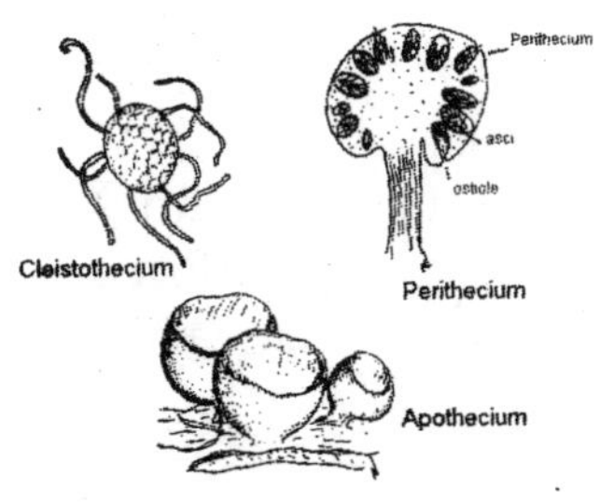

• **ascoconidium**
a budded spore arising from an ascospore within an ascus.

• **ascogenous hypha**
the restricted dikaryophase of many Ascomycetes, a dikaryotic hypha that grows out from the fertilised ascogonium.

• **ascogonium**
in Ascomycetes: the female gametangium, it may be unicellular or multicellular, simple or complex in form.

• **ascohymenial**
of, pertaining to, or having the characteristics of the Ascohymeniales.

• **ascohymeniales**
ascomycetes having asci (and paraphyses) developing as a hymenium and not in a preformed stroma.

• **ascolocular**
of, pertaining to, or having the characteristics of the Loculoascomycetes, the loculate Ascomycetes.

• **ascoma**
fungal fruiting body that contains ascospores, also termed an ascocarp.

• **ascomycete**
fungus of the subdivision Ascomycotina. In some taxonomic schemes these fungi form the class ascomycetes.

• **ascomycotina**
a subdivision of fungi characterised by the formation of sexually derived spores (ascospores) in asci. (ascomycetes, ascomycota)

• **ascophore**
a structure bearing asci (e.g., an ascus-producing hypha).

• **ascospore**
spores resulting from karyogamy and meiosis that are formed within an ascus. Sexual spore of the Ascomycota.

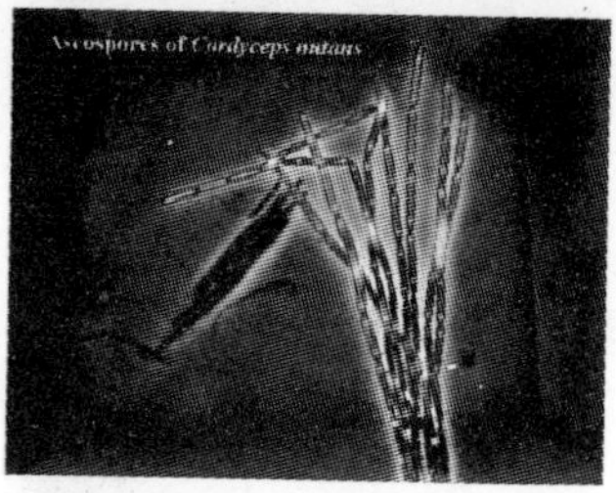

- **ascostroma**
a fructification of the Ascomycetes consisting of an undifferentiated mass of tissue or stroma on or in which the asci are developed (pl. ascostromata).

- **ascus**
a cell that is the site of meiosis and in which endogenous spores (usually meiospores) but sometimes also ascoconidia are formed.

- **aseptate**
lacking septa.

- **aseptic technique**
manipulating sterile instruments or culture media in such a way as to maintain sterility.

- **asexual**
this is a means of propagation that does not include seed production. Therefore propagating by cuttage, dividing and layering.

- **asexual propagation**
the duplication of a plant from a cell, tissue or organ of the plant.

- **asexual reproduction**
reproduction without fertilisation, especially relevant to cuttings and other types of vegetative reproduction.

- **aspergillus caesiellus**
this species is only occasionally pathogenic.

- **aspergillus carneus**
this species is only occasionally pathogenic.

- **aspergillus clavatus**
conidia dimensions 3-4.5 x 2.5-4.5 microns. Found in soils and animal manure. Can produce the toxin petulin, which may be associated with disease in hu-

mans and other animals. This species is only occasionally pathogenic.

- **aspergillus deflectus**
this species is only occasionally pathogenic.

- **aspergillus flavus**
conidia dimensions 3-6 microns or 3-5 microns. It grows on mouldy corn and peanuts. It can be found in warm soil, foods and dairy products. Some strains are capable of producing a group of mycotoxins - in the aflatoxin group. Aflatoxins are known animal carcinogen. There is limited evidence to suggest that this toxin is a human carcinogen. The toxin is poisonous to humans by ingestion. It may also result in occupational disease via inhalation. Experiments have indicated that it is teratogenic and mutagenic. It is toxic to the liver. It is reported to be allergenic. Its presence is associated with reports of asthma. It can be found in water damaged carpets. The production of the fungal toxin is dependent on the growth conditions and on the substrate used as a food source. This fungus is associated with aspergillosis of the lungs and disseminated aspergillosis. This fungus is occasionally identified as the cause of corneal, otomycotic and nasoorbital infections.

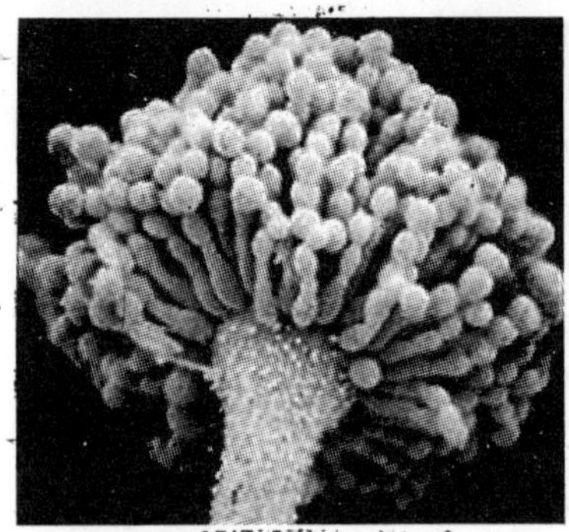

- **aspergillus fumigatus**
major cause of aspergillosis. Both invasive and allergic aspergillosis are caused by this organism. Aspergillosis affects individuals who are immune compromised. It is considered a human pathogen. It grows well at 35 degrees C. It is commonly found outdoors in compost piles with temperatures higher than 40 degrees C, in mild to warm soils and on cereals.

- **aspergillus glaucus**
  common outdoor fungus in the winter. It is reported to be allergenic. This species is only occasionally pathogenic. It can grow on leather. This fungus can grow at low moisture levels on grains, sugary food products, meat and wool. The ascomycetous state is Eurotium sp.

- **aspergillus nidulans**
  found in mild to warm soils and on slowly decaying plants. Can produce the mycotoxin sterigmatoc-ystin. This toxin has been shown to produce liver and kidney damage in lab animals. This fungus is associated with aspergillosis of the lungs and disseminated aspergillosis. This species is only occasionally pathogenic.

- **aspergillus niger**
  common cause of aspergillosis. It has a musty odour. It is commonly found in the environment on textiles, in soils, grains, fruits and vegetables . It has been reported to cause skin and pulmonary infections. It is a common cause of fungal related ear infections-otomycosis

- **aspergillus ochraceus**
  found in grains, soil and salted food products. It is not usually associated with decaying vegetation. Can produce a kidney toxin ochratoxin A which may produce ochratoxicosis in humans. This is also known as Balkan nephropathy. The toxin is produced at optimum growth conditions at 25 degrees C and high moisture conditions. The ochratoxin may also be produced by other Aspergillus sp. and Penicillium sp. Other toxins which can be produced by this fungus include penicillic acid, xanthomegnin and viomellein. These are all reported to be kidney and liver toxins.

- **aspergillus oryzae**
  this species is only occasionally pathogenic.

- **aspergillus parasiticus**
  some strains are capable of producing a group of mycotoxins - in the aflatoxin group. Aflatoxins are known animal carcinogen. There is limited evidence to suggest that this toxin is a human carcinogen. The toxin is a poisonous to humans by ingestion. Experiments have indicated that it is teratogenic and mutagenic. It is toxic to the liver. The production of the fungal toxin is dependent on the growth conditions and on the substrate used as a food source. Toxin production is dependent on the species or a strain within a species and on the food source for the fungus. Some of these toxins have been found to be carcinogenic in animal species. Several toxins are considered potential human carcinogens

- **aspergillus penicilloides**
  conidia dimensions 3-3.5 x 4-5 microns. Can grow in areas with low water activity. It is found in house dust and food.

- **aspergillus restrictus**
  this species is only occasionally pathogenic.

- **aspergillus**
  reported to be allergenic. Members of this genus are reported to cause ear infections. Many species produce mycotoxins, which may be associated with disease in humans and other animals.

- **aspergillus sydowi**
  this species is only occasionally pathogenic.

- **aspergillus versicolour**
  it is commonly found in soil, hay. cotton and dairy products, It can produce a mycotoxin sterigmatocystin and cyclopiaxonic acid. These toxins can cause diarrhea and upset stomach. It is reported to be a kidney and liver carcinogen. This species is only occasionally pathogenic.

- **asporogenous**
  not capable of forming spores

- **assimilation**
  1. the incorporation of inorganic forms of chemical elements into organic forms termed assimilates. Thus, we have the assimilation of CO_2 into photosynthetic such as

sugar and of NO_3, NH_4, and SO_4 into amino acids.
2. in growth analysis, net assimilation rate is given a specific definition as the rate of change of crop biomass (i.e., crop growth rate, reflecting the net of photosynthetic and respiratory activities) per unit leaf area index. Net assimilation rates can also be defined for carbon, nitrogen, or other entities.

- **assimilatory nitrate reduction**
  conversion of nitrate to reduced forms of nitrogen, generally ammonium, for the synthesis of amino acids and proteins.

- **associative symbiosis**
  close but relatively casual interaction between two dissimilar organisms or biological systems. The association may be mutually beneficial but is not required for accomplishment of a particular function.

- **atavism**
  recurrence of an ancestral character after an interval of several generations.

- **atrium**
  a structure that provides lots of light for plants. Commercial buildings often have their foyer as an atrium. Many homes have built in atriums.

- **auricle**
  a small lobe or ear.

- **autecology**
  the ecology of individual species, as opposed to that of communities.

- **autoecious**
  capable of completing a life cycle on one host.

- **autogamous**
  self-fertilised, flowers that are fertilised by their own pollen.

- **autolysis**
  spontaneous lysis.

- **autonomic movement**
  plant movement as a result of internal stimuli. Synonym: autogenic or spontaneous movement.

- **autoradiography**
  detecting radioactivity in a sample, such as a cell or gel, by placing it in contact with a photographic film.

• **autotrophic**
capable of growth independent of outside sources of nutrients or growth factors.

• **autotrophic nitrification**
oxidation of ammonium to nitrate through the combined action of two chemoautotrophic organisms, one forming nitrite from ammonium and the other oxidising nitrite to nitrate.

• **auxins**
a natural or artificial substance controlling the growth of shoots, roots, etc.

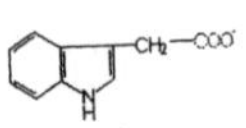
indol-3-acetic acid (IAA)

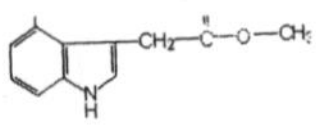
4-chloroindol-3-acetate

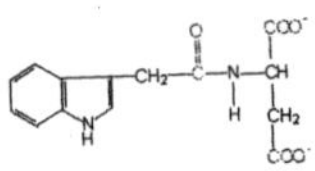
indol acetyl-aspartic acid

• **auxotroph**
a strain of micro-organism lacking the ability to synthesise one or more essential growth factors.

• **available pasture**
usage: not a recommended term (pasture refers to a specific type of grazing management unit, not to that which is consumed).

• **avirulent**
not exhibiting virulence, non-pathogenic.

• **avoidance**
1. ability of an organism to prevent an injurious stress, pathogen, or predator from penetrating its tissues (e.g., drought avoidance may be achieved through restriction of water loss or by expansion of the root system to a greater supply of water).
2. principle of plant disease control marked by deliberate actions to take advantage of environmental factors and time unfavourable for disease development

• **awl-shaped**
a narrow leaf which tapers to a stiff point.

• **axenic**
literally 'without strangers'. A system in which all biological populations are defined, such as a pure culture.

• **axial**
belonging to, around, in the direction of, or along an axis.

- **axil**

  the angle between the upper surface of a leaf or leaf stalk and the stem that carries it. A growth or flower bud ('axillary bud') often appears in the axil.

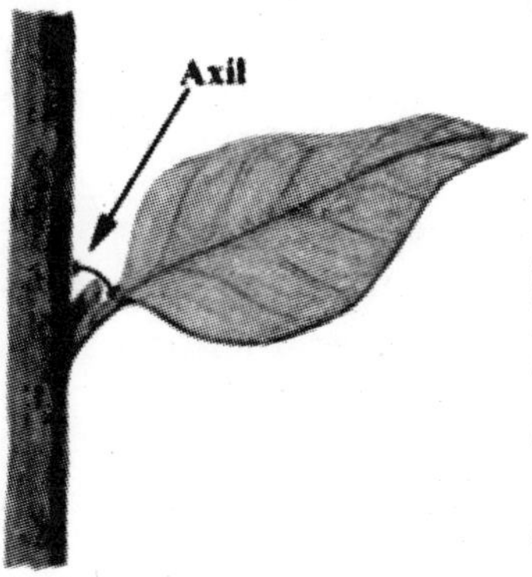

- **axis**

  1. upper angle formed between the stem or branch and any other branch, leaf or other organ arising from them.

  2. the main line of growth in a plant or organ, e.g., the stem, from which the other parts such as the leaves and flowers grow.

- **azygospore**

  a parthenogenetic zygospore, formed by some vesicular-arbuscular mycorrhizal fungi.

- **b and b**

  balled and burlap, a method in which plants are sold where the roots of a plant have been lifted and wrapped in burlap (sometimes plastic covered material) to keep it together until transplanted. Large trees are often sold this way.

- **bacillar**

  shaped like a short rod.

- **bacilliform**

  rod-shaped

- **bacillus**

  1. a member of the genus Bacillus.

  2. any rod-shaped bacterial cell, i.e., a cell whose length is two or more times greater than its width.

- **backbulb**

  refers to the propagation of an orchid. It is the old, dormant pseudobulb. It may be leafless, but will still produce a new plant.

- **backcross**
to cross (mate with one parent).

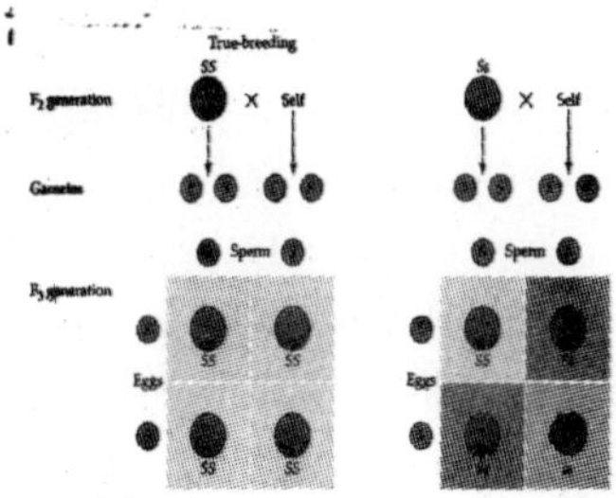

An F_2 Generation

- **backfill**
replacing dirt from the original hole after planting.

- **backgrounding**
intensive management of young cattle (post-weaning) to facilitate maximum performance.

- **backyard wildlife habitat**
a dreamy situation where native plant materials are providing food and shelter for protection and reproduction for birds, insects, and mammals in one's own backyard.

- **bacteria**
microscopic organisms having round, rod-like, spiral or filamentous single-celled or no cellular bodies often gathered into colonies.

- **bactericidal**
lethal to bacteria.

- **bacteriochlorophyll**
light-absorbing pigment found in green sulphur and purple sulphur bacteria.

- **bacteriocin**
a protein antibiotic, one or more types of which can be produced and 'exported' (excreted by certain strains of bacteria).

- **bacteriophage**
virus that infects bacteria, often with destruction or lysis of the host cell.

- **bacteriostatic**
able to inhibit the growth and reproduction of at least some types of bacteria.

- **bacteroid**
altered form of cells of certain bacteria. Refers particularly to the swollen, irregular vacuolated cells of rhizobia in nodules of legumes.

- **balanced diallelic**
the genotype involving a multiple allelic locus in an auto-tetraploid where two different alleles are represented an equal number of times.

• **bale**
the area of the trunk between the base of a tree and the lowest branch.

• **ballistospore**
a spore that when mature is actively projected

• **band fertilise**
to apply fertiliser in a narrow line along a row of plants or in a circle around individual plants.

• **barbate**
bearded.

• **bare fallow**
complete inversion and incorporation of residues for maximum decomposition, done to prevent the growth of all vegetation, usually associated with summer fallow.

• **bare root**
plants that have been dug out of the ground when dormant. The soil is shaken free, washed and stored until shipment. Roses and daylilies commonly come this way, as well as smaller shrubs and bushes, sold in their dormancy.

• **bark ringing**
taking of a circle to bark from a fruit tree to inhibit growth but to encourage cropping.

• **basal**
at the base of an organ or part such as the stem or pseudo-bulb.

• **base composition**
proportion of the total bases consisting of guanine plus cytosine or thymine plus adenine base pairs. Usually expressed as a guanine + cytosine (G + C) value, e.g., 60% G+C.

• **base dressing**
fertiliser or compost put on or dug in to the soil before planting or sowing.

• **basidiocarp**
a sporocarp produced by a member of the Basidiom-

ycotina and which bears basidiospores.

- **basidiole**
a structure in the hymenium of a member of the Basidiomycotina that is morphologically similar to a basidium without sterigmata. It may be an immature basidium or a permanently sterile structure in the hymenium.

- **basidioma**
fruiting body that produces basidia, also termed a basidiocarp.

- **basidiomycetes**
fungal spores which are from mushrooms. The specific mushroom species can not be identified on the culture plate. Many mushroom spores are reported to be allergenic.

- **basidiomycota**
see **basidiomycotina**.

- **basidiomycotina**
a subdivision of fungi characterised by the formation of basidiospores on basidia.

- **basidiospore**
spore resulting from karyogamy and meiosis that are formed on a basidium.. Sexual spore of the Basidiomycota.

- **basidium**
a cell in which karyogamy and meiosis take place and which bears exogenous spores of sexual origin. (pl. basidia)

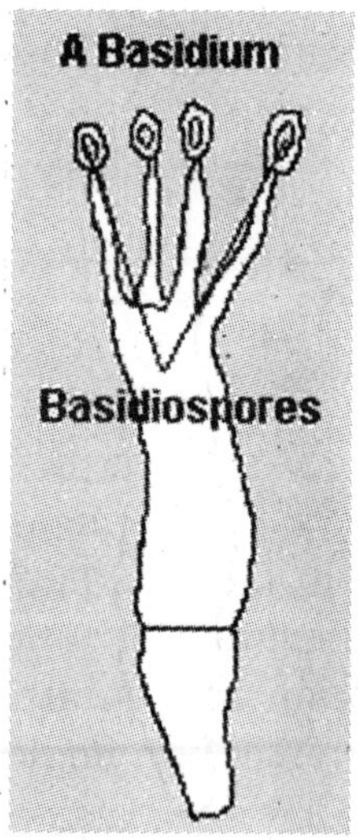

- **basipetal**
successive from apex to base.

• **batch drying**
drying seed in relatively small quantities held in a stationary position, as opposed to drying in a continuous moving line.

• **bearded**
a petal bearing a tuft or row of long hairs.

• **bedding plant**
usually an annual plant temporarily in a garden display. Sometimes called carpet bedding.

• **bell jar**
an old term and glass container which is bell shaped. On the top is a knob for the use of protection of a delicate plant. From the Victorian Era and now-a-days known as a cloche.

• **bench setting**
height that the cutting plane of a mower is set above a hard, level surface.

• **beneficial insect**
these are insects that will improve and work in our gardens. By improving the soil, going after harmful insects, and pollinating plants. Ladybeetles, earthworms, and bees are well known.

• **bent grass**
a high-maintenance grass used on putting greens. It requires frequent cutting with a reel mower, frequent fertilisation and watering. It is highly susceptible to several diseases.

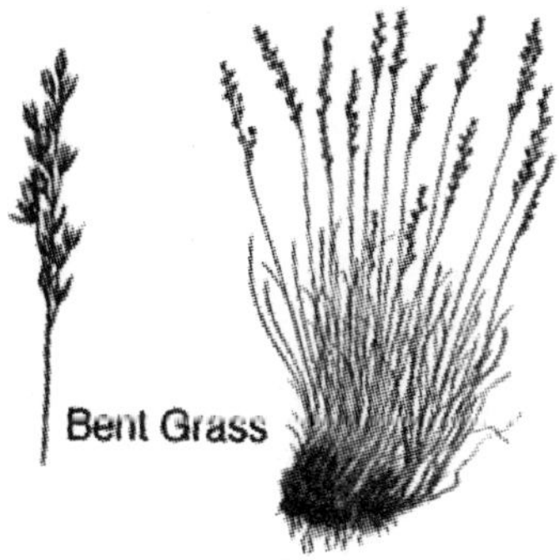

• **berm**
a landscaping technique that is used to create interest, privacy, or screening. It may also divert water runoff.

Creating a mound of earth or a hill makes it.

- **bicollateral vascular bundle**
  a vascular bundle with phloem on both sides of the xylem instead of on the outer side only.

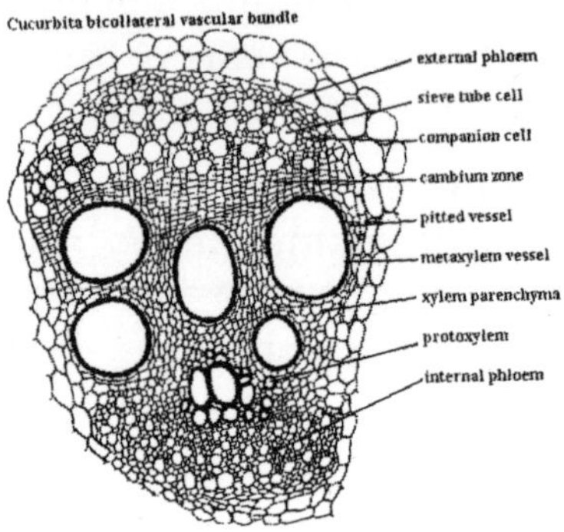

- **bicolour**
  a flower with petals which bear two distinctly different colours.

- **biennial**
  a plant that will require two growing seasons to complete its life cycle. In the first year it has leaves. In the second year it has blooms and seeds (e.g. foxglove, hollyhock, rose campion).

- **bifoliate**
  with two leaves.

- **bifurcate**
  dividing into two branches.

- **bigeneric**
  a hybrid produced by crossing two different genera.

- **binary fission**
  fission in which two cells, usually of similar size and shape, are formed by the growth and division of one cell.

- **binomial nomenclature**
  a system in which the scientific name of a plant consists of two parts indicating the genus and species.

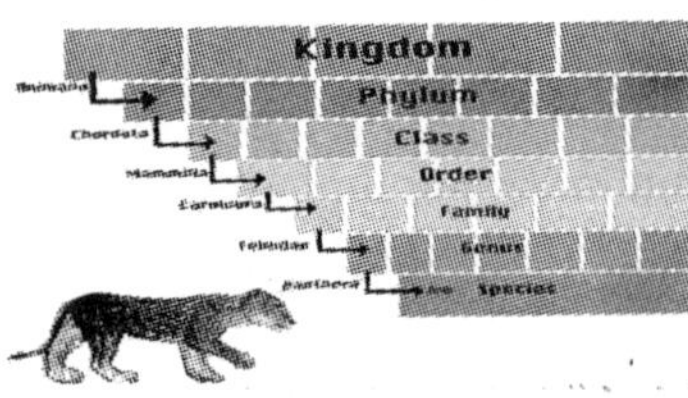

- **bioaccumulation**
  accumulation of a chemical substance in living tissue.

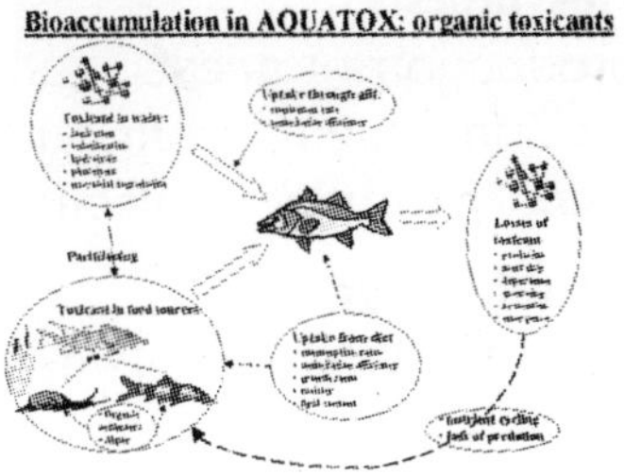

- **bioassay**
  1. the use of living organisms to quantitatively estimate the amount of biologically active substances present in a sample.
  2. in cell biology and molecular genetics: determination of the effectiveness of a compound by measuring its effect on plants or animal tissues or organisms in comparison with a standard preparation.

- **Biochemical Oxygen Demand (BOD)**
  amount of dissolved oxygen consumed in five days by biological processes breaking down organic matter.

- **biocontrol**
  see **biological control**.

- **biodegradable**
  substance capable of being decomposed by biological processes.

- **biofilm**
  microbial cells encased in an adhesive, usually a polysaccharide material, and attached to a surface.

- **biogeochemistry**
  study of microbially mediated chemical transformations of geochemical interest, such as nitrogen or sulphur cycling.

- **biological control**
  the deliberate use by humans of one species of organism to eliminate or control another.

- **biological diversity**
  presence of many different types of living organisms.

- **biological pest control**
  using living organisms such as beneficial insects or parasites to destroy garden pests.

- **biomagnification**
  increase in the concentration of a chemical substance as it progresses to higher trophic levels of a food chain.

- **biomass**
  the weight of living organisms (plants and animals) in an ecosystem, at a given

point in time, expressed as fresh or dry weight.

- **bioremediation**
use of micro-organisms to remove or detoxify toxic or unwanted chemicals from an environment.

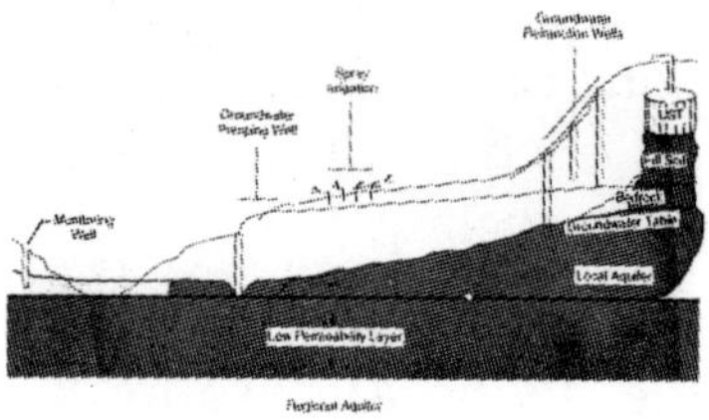

- **biosolid**
the resides of wastewater treatment. Formerly called sewage sludge.

- **biosphere**
zone incorporating all forms of life on earth. The biosphere extends from deep in sediment below the ocean to several thousand meters elevation in high mountains.

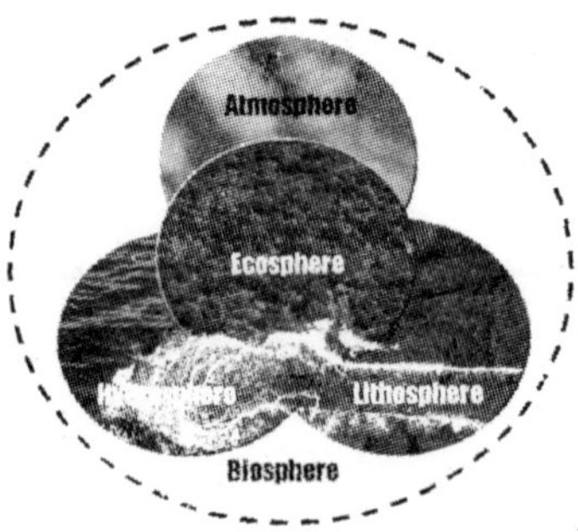

- **biosynthesis**
production of needed cellular constituents from other, usually simpler, molecules.

- **biotechnology**
use of living organisms to carry out defined physio-chemical processes having industrial or other practical application.

- **biotroph**
an organism that derives nutrients from the living tissues of another organism (its host).

- **biotype**
a subspecies of organism morphologically similar to but physiologically different from other members of the species.

- **bipolar**
at both ends or poles.

- **bipolaris**
a fungus with large spores which would be expected to be deposited in the upper

respiratory tract. This fungus can produce the mycotoxin - sterigmatocystin- which has been shown to produce liver and kidney damage when ingested by laboratory animals.

- **bird resistance**
  a characteristic of a genotype or individual plant in which it is avoided by birds until other food sources are exhausted, or until the plant is weathered.

- **biseptate**
  having two septa.

- **bisexual**
  two-sexed, with both stamens and pistils.

- **bitunicate**
  having two walls.

- **blade**
  the expanded part of a leaf or petal.

- **blanch**
  to keep light from the leaves and stems, keeping the plant tissue soft (e.g., endive is grown this way).

- **blastic**
  one of two basic kinds of conidiogenesis. There is a marked enlargement of a recognisable conidium before it is delimited by a [illegible]tum.

- **blasting**
  a symptom of plant disease characterised by shedding of unopened buds, classically, the failure to produce fruit or seed

- **blastoconidium**
  see **blastospore**.

- **blastomyces**
  human pathogen. The fungus is commonly found in soil. It is a dimorphic fungus which has filamentous fungus when grown at 25 degrees C and a yeast form at 37 degrees C.

- **blastospore**
  a spore that arises by budding, as in yeasts.

- **bleed**
  to lose sap through a cut or wound.

- **blend, seed**
  a combination of seeds of varieties and composite-cross populations of the same kind, each present in excess of 5% by weight of the whole, and derived by growing the component varieties and/or composite-cross

populations separately and mixing seed to form the commercial class of seed.

- **blight**
a disease characterised by rapid and extensive death of plant foliage. A general term applied to any of a wide range of unrelated plant diseases e.g. chestnut blight, fire blight, late blight, halo blight.

- **blind**
the loss of the growing point, resulting in stoppage of growth. Also, failure to produce flowers or fruit.

- **bloom**
a natural mealy or waxy coating covering the leaves of some house plants.

- **blooming**
in the grass family: the period during which florets are open and anthers are extended. Synonym: anthesis.

- **blossom end rot**
a cultural deficiency created by a lack of calcium. Very closely related to extreme temperatures, uneven watering, and root damage. Most commonly seen in peppers and tomatoes.

- **blot**
see **blotting**.

- **blotch**
a disease characterised by large, and irregular in shape, spots or blots on leaves, shoots, and stems.

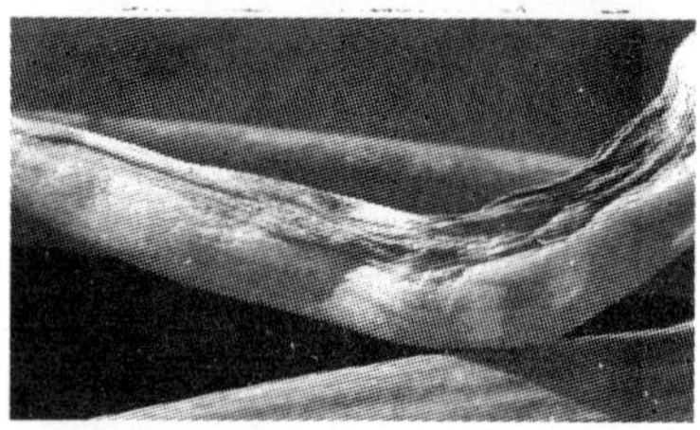

- **blotting**
following electrophoresis: the transfer of nucleic acids and/or proteins from a gel strip to a specialised, chemically reactive matrix on which the nucleic acids, etc. may become covalently bound in a pattern similar to that present in the original gel.

• **bog plant**
plants that prefer damp and moist soil as their habitat (e.g. pitcher plant, Venus fly trap.)

• **bole**
the area of a tree trunk that is from the ground to the first major branch.

• **bollsize**
weight in grams of seed cotton from one boll.

• **bolt**
annual vegetables or flowers that grow quickly to flowering stage, at the expense of their best overall development, and go to seed (e.g., dill in hot weather).

• **bonemeal**
a fertiliser made from crushed animal bones. It is a natural high phosphorus fertiliser, very slow releasing and good for root development.

• **bones-of-the-garden**
in the language of garden design this is the permanent structural elements that give the shape to gardens: paths, walls, steps, fences, trellises, seats, water gardens, and hedges.

• **bonsai**
refers to either the trees or the method of producing miniature trees by pruning roots and restricting the size of the container. Originated in Japan.

• **boss**
a ring of prominent and decorative stamens.

• **botanical name**
the Latin scientific name of a plant is its botanical name. There is only one botanical name per plant so if you

want a specific variety, use its botanical name to be sure you are getting what you want. Common names tend to be confusing.

- **botanical variety**

an infraspecific taxon in botanical nomenclature, below the rank of subspecies. Abbr.: var.

- **botrytis**

conidia dimensions 7-14 x 5-9 microns. Reported to be allergenic. It is parasitic on plants and soft fruits. Found in soil and vegetables. Possibly associated with allergic symptoms (skin tests).

- **bottle garden**

a form of terrarium in which a large and heavy glass container such as a carboy is used.

- **bottom heat**

an arrangement used in propagation. Usually electric heating cables will run through the base of the propagation medium. Great for seed germination and cuttings.

- **bract**

a modified leaf, often highly coloured and sometimes mistaken for a petal. Examples of house plants with showy bracts are Poinsettia, Aphelandra and Bougainvillea.

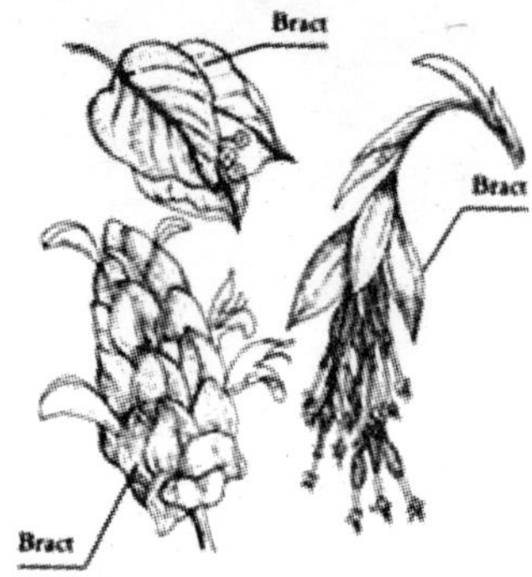

- **bracteate**

bearing bracts

- **bramble**

any shrub with thorns in the rose family, usually refers to blackberries and raspberries.

- **bran**
  pericarp of grain.
- **branch crown**
  plant tissue that is the junction of the roots and stem that forms on the side of a strawberry plant. These only form foliage.
- **break**
  production of a side shoot after removal of the growing point.
- **breaking load**
  the maximum force applied to a structure in a tensile test carried to rupture.
- **broad spectrum**
  pesticides that affect a wide variety of pests.
- **broadcast**
  a method by which seeds or fertiliser are scattered randomly to cover an area.
- **broadcast application**
  application by spreading or scattering on the soil surface.
- **broadcast sprigging**
  vegetative turf establishment by broadcasting and covering of stolons, rhizomes, or tillers with soil.
- **broom**
  in plant pathology: a symptom in which lateral branches proliferate in a dense cluster on the main branch (witches'-broom).
- **broth**
  in microbiology: any of a variety of liquid media, especially nutrient broth or any liquid medium based on nutrient broth and/or hydrolysed protein.
- **brown rot**
  a fungus that is a very common disease on fruit. Buy disease resistant varieties. Remove all infected parts of the plant.

- **brown rot fungus**
  fungus that attacks cellulose and hemi cellulose in wood, leaving dark-coloured lignin and phenolic materials behind.
- **browning**
  the reaction between reducing sugars and free amino

groups in proteins to form a complex that undergoes a series of reactions to produce brown polymers usually referred to as melanoidins. Higher temperatures and basic pH favour the reaction. The process is desirable for flavour and aroma in certain food processing situations (e.g., toasting bread) but is detrimental in other situations (e.g., drying seed and food products).

- **browse**

  1. leaf and twig growth of shrubs, woody vines, trees, cacti, and other non-herbaceous vegetation available for animal consumption.

  2. to consume browse in situ (of animals).

- **BT**

  Bacillus Thuringiensis a bacterium which will destroy the stomach cells of insects that consume it. It degrades quickly in sunlight so spray early in the evening. This biological insecticides will also kill young butterfly caterpillars.

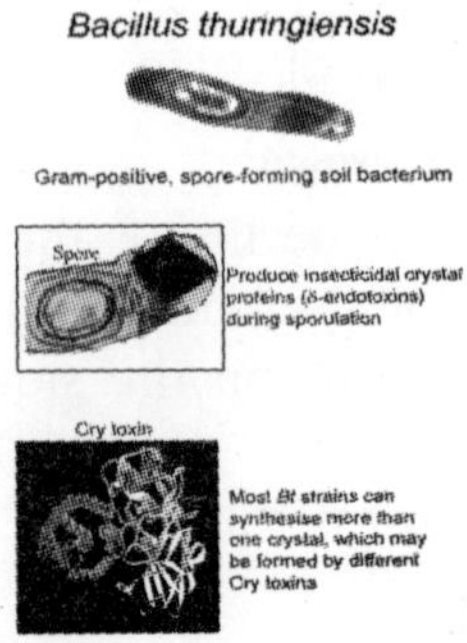

- **buccal capsule**

  in nematodes: structure connecting the oral opening with the anterior portion of the oesophagus. The buccal capsule (also called the stoma) is subject to great variation among different nematodes.

- **bud**

  an unopened flower.

- **bud scales**

  specialised tissue that covers the terminal bud and embryonic leaves of a plant during winter.

- **bud union**
  the location of a graft.
- **budding**
  asexual reproduction (usually for yeast) beginning as a protuberance from the parent cell that grows and detaches to form a smaller, daughter cell.
- **bulb**
  this is really a modified stem acting as storage for the plant (that's why you shouldn't cut the foliage back on daffodils when they stop flowering because it feeds the bulb). A true bulb is made up of individual, although often tightly packed,

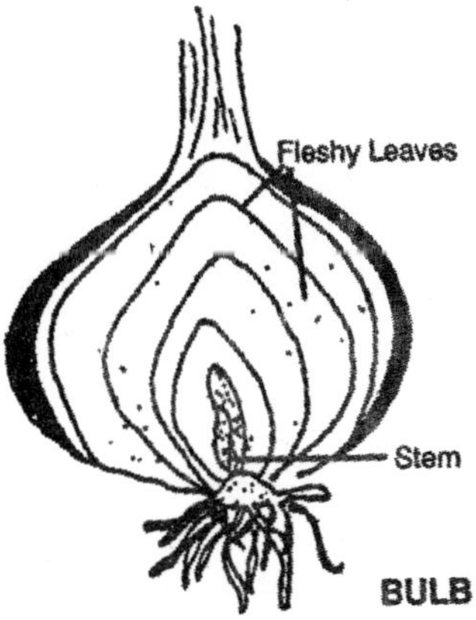

BULB

  scales. The term bulb is also applied wrongly to tubers and rhizomes.
- **bulb fibre**
  special compost formulated for growing bulbs like hyacinths and narcissus.
- **bulbil**
  a small bulb that forms along the stems of certain plants, such as tiger lilies and bladder ferns.
- **bulblet**
  an immature small bulb formed at the base of a mature bulb, e.g Hyacinth.
- **bulbous**
  bulb-like.
- **bulk density, soil**
  mass of dry soil per unit bulk volume (combined volume of soil solids and pore space).
- **bullae**
  in nematodes: blisterlike prominences near the vulval fenestra of some Heteroderidae.
- **bursa**
  in nematodes: caudal alae of males used to clasp the female during copulation.
- **bursicle**
  a membranous pocket or pouch in the orchid flower, covering or enclosing the

viscidium to stop it from drying up, and being pushed back by visiting insects.

- **bush**
  a many branched small shrub with no distinct main stems.

- **button**
  the small heads of broccoli or cabbage that form as a result of seedlings being exposed to freezing temperatures.

- **caducous**
  falling off at early stages, when buds fall off.

- **caeoma**
  an aecium in the uredinales that is not surrounded by a peridium, from the form genus Caeoma. (pl. caeomata.)

- **caespitose**
  growing in tufts or dense clumps.

- **cage**
  an enclosure used to support a plant.

- **calcareous**
  containing calcium carbornate, or calcite Chalky.

- **calceolate**
  slipper-like, with the form of a round-toed shoe.

- **calcicole**
  a plant that thrives on alkaline soil.

- **calcifuge**
  a plant that hates alkaline soils and will not grow in them.

- **calcined clay**
  clay minerals, such as montmorillonite and attapulgite, that have been fired at high temperatures to obtain absorbent, stable, granular particles, used as amendments in soil modification.

• **calcitic limestone**
a common material used for 'liming' soil that has an acid level that is too high. This type is most commonly used and contains calcium carbonate.

• **caliche**
Caliche is a deposit of calcium carbonate (lime) beneath the soil surface. This condition is more commonly called 'hardpan' and creates an impervious layer in lower levels of soil.

• **callus**
scar tissue that forms when a plant has been damaged or cut. When propagating some succulents it is best to have the leaf form a callus, to prevent disease and rotting.

• **calvin cycle**
biochemical route of carbon dioxide fixation in many autotrophic organisms.

• **calyx**
the calyx is formed by the sepals of a flower and describes the green parts that cover the flower bud before it opens and then remain at the base of the flower afterwards.

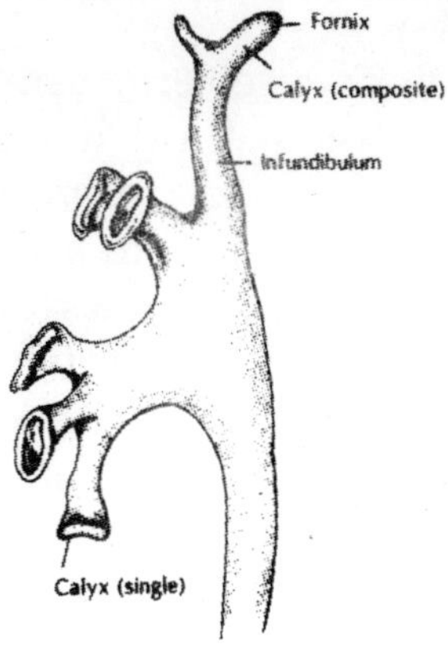

• **cambium**
this is the thin membrane that grows just under the bark of a plant.

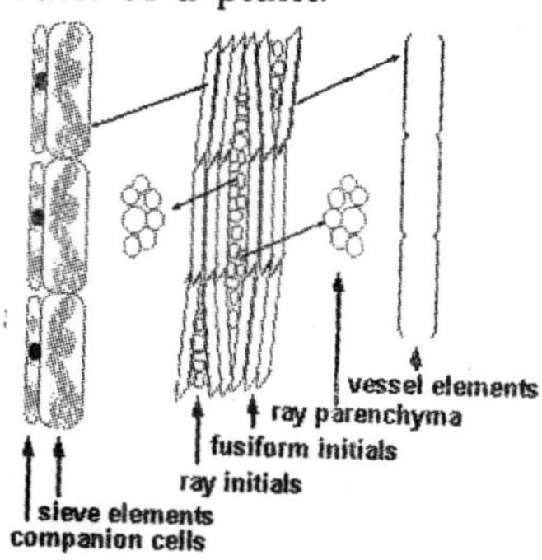

• **candle**
the new shoot growth on needled evergreens before the needles expand.

• **cane**
a slender, straight, not very woody branch or stem of a plant (e.g., bamboo, rose, raspberry and blackberry bushes.).

- **canker**
an imprecise term usually used for a plant disease characterised (in woody plants) by the death of cambium tissue and resulting loss and malformation of bark, or (in non-woody plants) by the formation of sharply delineated, dry, necrotic, localised lesions on the stem. The term 'canker' may also be used to refer to the lesion itself, particularly in woody plants.

- **canopy**
the vertical projection downward of the aerial portion of plants, usually expressed as percent of ground so occupied.

- **capillary action**
the natural upward movement of water in confined areas, such as the spaces between soil particles.

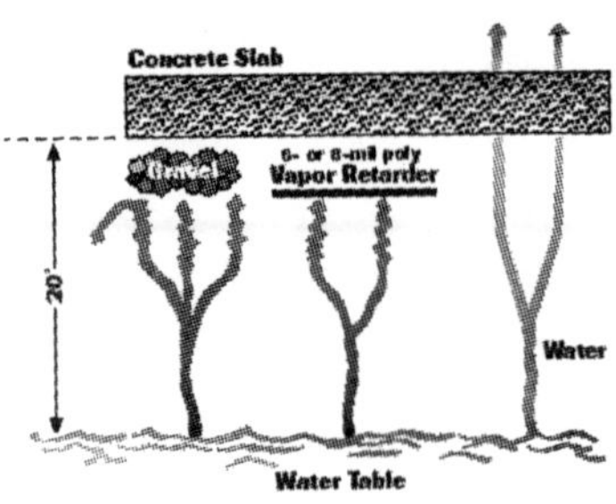

- **capillitium**
a mass of sterile fibres interspersed among spores within a sporocarp (in the Gasteromycetes and Myxomycota) (pl. capillitia).

- **capitulum**
in nematodes: medial ventral sclerotisation of the spicular pouch.

- **cappillary matting**
special matting to put on benches under pots and seed trays. The matting allows water to be drawn up into the soil.

- **capsid**
the protein shell that surrounds the virus nucleic acid.

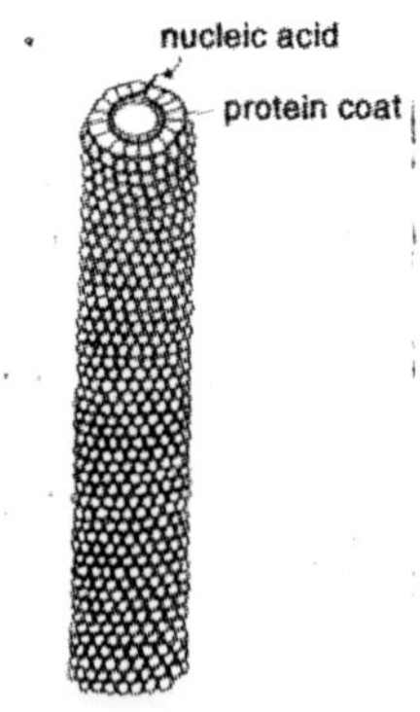

- **capsule**
1. a dry fruit which opens, when the seeds are ripe, at several slits or holes. Any

closed vessel containing spores or seeds.

2. compact layer of polysaccharide exterior to the cell wall in some bacteria.

- **carbohydrates, structural**
carbohydrates found in the cell walls (e.g., hemi cellulose, cellulose).

- **carbon cycle**
sequence where carbon dioxide is converted to organic forms by photosynthesis or chemo-synthesis, recycled through the biosphere, with partial incorporation into sediments, and ultimately returned to its original state through respiration or combustion.

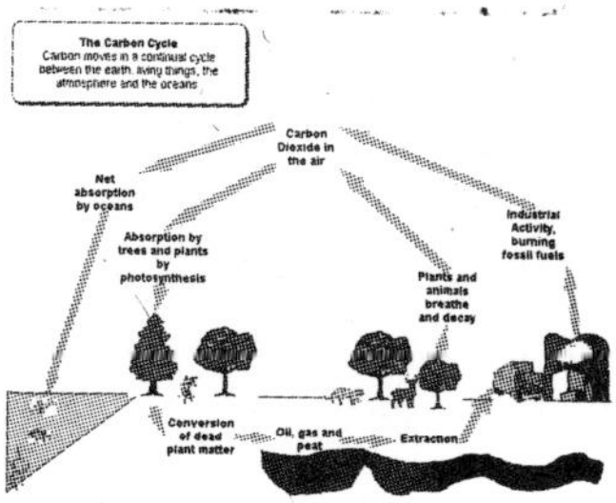

- **carbon dioxide, CO^2**
a colourless, odourless gas found in the air. It is absorbed by plants and exhaled by animals.

- **carbon fixation**
conversion of carbon dioxide or other single-carbon compounds to organic forms such as carbohydrates.

- **carbon-nitrogen (C/N) ratio**
ratio of the mass of organic carbon to the mass of nitrogen in soil or organic material.

- **carboxyl group**
a COOH group attached to a carbon skeleton as in the carboxylic acids and fatty acids.

- **carboy**
a large and heavy glass vessel, originally designed for the storage of chemicals but now commonly used as a container for bottle gardens.

- **carcinogen**
substance which causes the initiation of tumour formation. Frequently a mutagen.

- **cardia**
in nematodes: valvular apparatus connecting the oesophagus and intestine. Sometimes called the cardiac valve or oesophago-intestinal valve.

- **carding**
  the process of untangling and partially straightening fibres by passing them between two closely spaced surfaces, at least one of which is covered with sharp points, that move at different speeds, thus converting a tangled mass of fibres to a filmy web.

- **carlavirus**
  siglum of carnation latent virus. Member of a group of plant viruses with slightly flexuous, rod-shaped particles containing a single molecule of linear RNA, most of which are transmitted by aphids in a non-circulative manner.

- **carmovirus**
  siglum of carnation mottle virus. Member of a group of plant viruses with small, isometric particles containing a single molecule of linear RNA, transmitted in nature through soil and (rarely) by an insect vector.

- **carnivorous**
  used in the gardening world to denote a plant (usually tropical) that typically lives in highly acidic soil that doesn't adequately provide enough nourishment. Nature has adapted these plants to trap and consume insects for this need. An example is the Venus Flytrap plant.

- **carpel**
  simple pistil, or one member of a compound pistil, spore bearing organ.

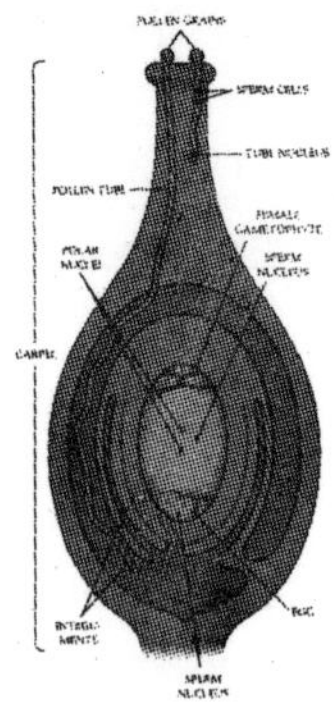

- **carrier**
  an organism that bears an infectious agent internally but shows no marked symptoms of the disease caused by that agent.

- **carrying capacity**
  the maximum stocking rate that will achieve a target level of animal performance in a specified grazing method, that can be applied over a defined period of time

without deterioration of the ecosystem. Usage: carrying capacity is not static from season to season or year to year and may be defined over fractional parts of years. Average carrying capacity refers to long-term carrying capacity averaged over years, annual carrying capacity refers to a specific year.

- **castings, earthworm**
  soil and plant remains excreted and deposited by earthworms in or on the turf surface or in their burrows. Castings form relatively stable soil granules that can, on closely mowed turf, produce an uneven surface. Synonym: earthworm casts.

- **catabolism**
  biochemical processes involved in the breakdown of organic compounds, usually leading to the production of energy.

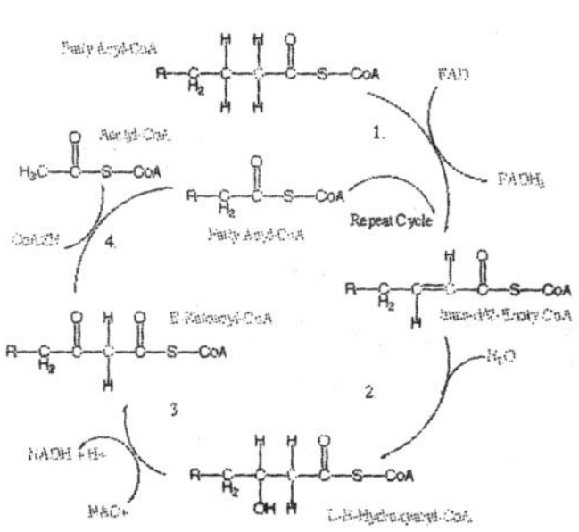

- **catabolite repression**
  transcription-level inhibition of a variety of inducible enzymes by glucose or other readily used carbon source.

- **catalyst**
  substance that promotes a chemical reaction by lowering the activation energy without itself being changed in the end. Enzymes are a type of catalyst.

- **caterpillar**
  worm-like larva of various insects, especially butterflies and moths.

- **Cation Exchange Capacity (CEC)**
  sum of exchangeable cations that a soil can adsorb at a specific pH. Expressed as centimoles of positive charge per kilogram of soil (cmolc kg^{-1}).

- **catkin**
  usually petal-less flowers arranged in a spike.

- **caudal**
  in nematodes: Pertaining to or located near the posterior region or tail.
- **caudate**
  having a 'tail' or narrowed, apical extension, like some sepals and petals.
- **caudicle**
  slender stalk-like appendage that attaches the viscidium (a sticky gland) to the pollinia (pollen packets).
- **causal agent of disease**
  that which is capable of causing disease
- **C-banding**
  Centromere Banding.
  1. technique for chromosome identification, using modified Giemsa staining to reveal the distribution of constitutive heterochromatin, expressed at principal stages of meiotic and mitotic division.
  2. the banding pattern thus revealed.
- **cell**
  the unit of plants that makes up tissues. Cells have a cell wall that encloses the protoplasm.
- **cell cycle**
  the period from one cell division to the next.
- **cell membrane**
  see **cytoplasmic membrane**.
- **cell pack**
  a group of gardeners travelling together in a confined space for snipping and stealing plant material in a botanical garden.
- **cell wall**
  layer or structure that lies outside the cytoplasmic membrane. It supports and protects the membrane and gives the cell shape.
- **cellulose**
  glucose polysaccharide (with beta-1,4-linkage) that is the main component of plant cell walls.

Cellulose

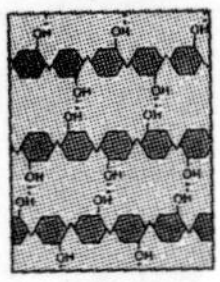

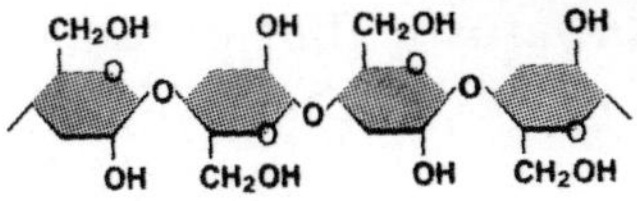

- **cell-wall constituents**
compounds that make up or constitute the cell wall, including cellulose, hemi cellulose, lignin, and minerals (ash), i.e., neutral-detergent fibre.

- **centre of origin**
the geographic region in which a taxon (species, genus, family, etc.) or a crop originates.

- **centrifugal spreader**
an applicator from which dry particulate material is broadcast as it drops onto a spinning disk or blade beneath the hopper.

- **cephalic**
in nematodes: pertaining to or located near the head.

- **cephalids**
in nematodes: two structures (posterior and anterior) situated in the cephalic region and extending in a complete circle around the body, possibly part of the nervous system.

- **certified seed**
certified seed is the progeny of breeder, select, foundation, or registered seed so handled as to maintain satisfactory genetic purity and identity, and which has been acceptable to the certifying agency. Certified tree seed is defined as seed from trees of proven genetic superiority, as defined by the certifying agency, produced so as to assure genetic identity. Seed from inter-specific hybrids or forest trees may be included.

- **CFU**
Colony-Forming Unit.

- **chaetomium**
large ascomycetous fungus producing perithecia. It is found on a variety of substrates containing cellulose including paper and plant compost. It has been found on paper in sheetrock. It is reported to be allergenic. Can produce an Acremonium like state on fungal media.

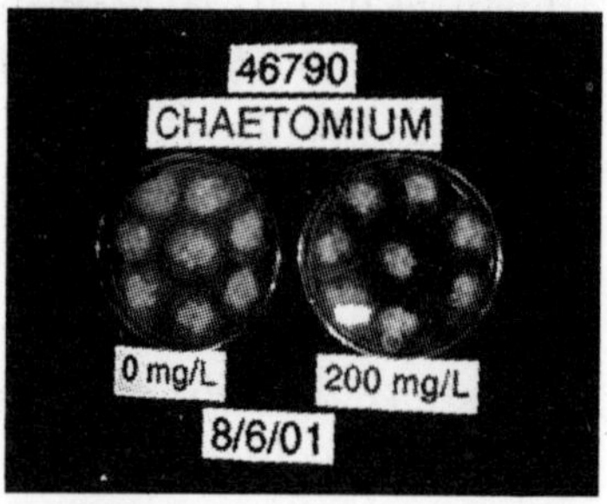

- **chalky**
  colour descriptor (related to hardness) characterising kernel endosperm of cereal grains, especially rice. A rice kernel with a high level of chalk is generally undesirable. The chalky appearance arises from the structure of the endosperm, which has many small voids between the starch granules and the protein matrix. These voids cause light to be refracted and hence the endosperm appears white to reflected light and opaque to transmitted light. Synonyms: floury, soft endosperm.

- **chelate (chelator)**
  organic chemical that forms ring compound in which a metal is held between two or more atoms strongly enough to diminish the rate at which it becomes fixed by soil, thereby making it more available for plant and microbial uptake.

- **chemoautotroph**
  organism that obtains energy from the oxidation of chemical, generally inorganic, compounds and carbon from carbon dioxide.

- **chemoheterotroph**
  organism that obtains energy and carbon from the oxidation of organic compounds.

- **chemolithotroph**
  organism that obtains energy from the oxidation of inorganic compounds and uses inorganic compounds as electron donors.

- **chemoorganotroph**
  organism that obtains energy and electrons (reducing power) from the oxidation of organic compounds.

- **chemostat**
  continuous culture device usually controlled by the concentration of limiting nutrient and dilution rate.

- **chemotaxis**
  oriented movement of a motile organism with reference to a chemical

agent. May be positive (toward) or negative (away) with respect to the chemical gradient.

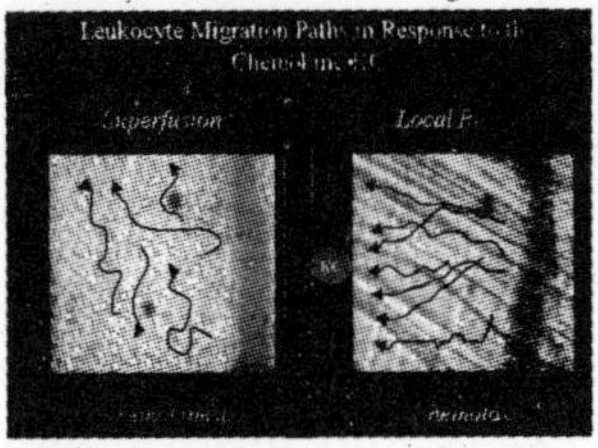

- **chemotherapy**
  the use of chemical(s) (e.g., antibiotics or fungicides) for the treatment of a disease.

- **chilling injury**
  damage to plants at low temperatures in the absence of freezing. Chilling injury is common in plants of tropical or subtropical origin at temperatures less than 10 °C (50 °F) or, for some fruits, less than 15 °C (59 °F). The cause usually is a change in viscosity of lipids in membranes.

- **chlamydospore**
  thick-walled resting structure that forms from the cell wall of a fungal hypha, usually formed under conditions where the hypha is no longer able to function optimally.

- **chloranemia**
  the necrotic symptom of yellowing, a loss of chlorophyll.

- **chlorophyll**
  the green pigment in the leaves and sometimes stems of most plants, which uses solar energy to convert carbon dioxide and water to sugar.

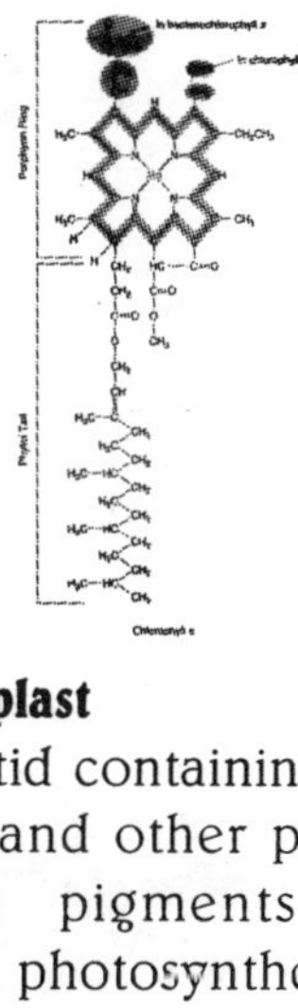

- **chloroplast**
  a plastid containing chlorophyll and other photosynthetic pigments within which photosynthesis takes place.

- **chlorosis**
  the loss of chlorophyll from the tissues of a plant, resulting from microbial infection, viral infection, the action of certain phytotoxins, the lack of light, to magnesium or iron deficiency, etc. Chlo-

rotic tissues commonly appear yellowish.

- **chlorotic**
  see **chlorosis**.

- **chord**
  in nematodes: a longitudinal internal thickening of the hypodermis.

- **chromatography**
  any technique used to separate different species of molecules (or ions) by subjecting them to two different carrier phases: mobile and stationary phases.

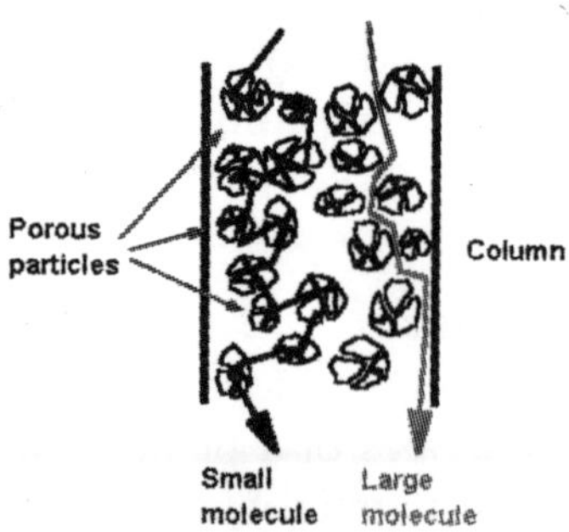

- **chromic oxide**
  Cr_2O_3. A completely indigestible chemical used as an indicator to estimate forage intake.

- **chromosome**
  genetic element carrying information essential to cellular metabolism. Prokaryotes have a single chromosome, consisting of a circular DNA molecule. Eukaryotes contain more than one chromosome, each containing a linear DNA molecule complexed with specific proteins.

- **chytrid**
  fungal organism in the phylum Chytridiomycota that consists of a spherical cell from which short thin filamentous branches (rhizoid) grow that resemble fine roots.

- **ciliate**
  protozoan that moves by means of cilia on the surface of the cell.

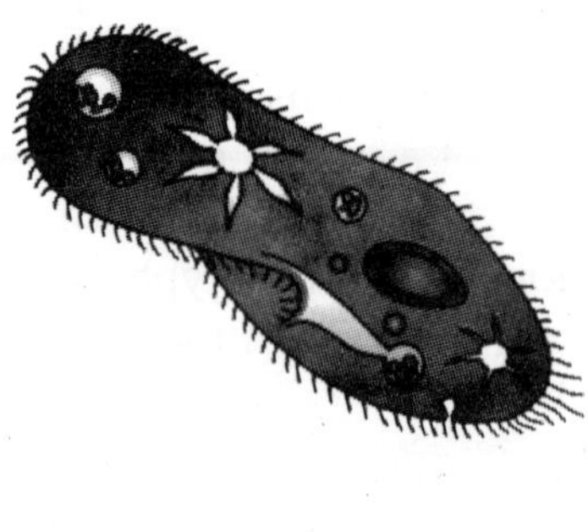

- **cilium (cilia)**
  short, threadlike appendages that extend from the surface of some protozoa and beat rhythmically to propel them.

- **circulative transmission**
  virus transmission characterised by a long period of acquisition of the virus by a vector, a latent period of several hours before the vector is able to transmit the virus, and retention of the virus by the vector for a long period, usually several days. Also termed Persistent Transmission.

- **cirrhus (cirrus)**
  1. a mass of spores in the form of a ribbon or tendril, forced from the fruiting body of a fungus.
  2. a discrete group of somatic cilia (several to over 100) that act primarily as a unified locomotive organelle. The typical cirrus is conical (pl. cirrhi or cirri.)

- **citric acid cycle**
  see **tca cycle**.

- **cladode**
  a modified stem which has taken on the form of a leaf, e.g. the needle like 'leaves' of Asparagus Fern.

- **cladosporium fulvum (fulvia fulva)**
  conidia dimensions 12-47 x 4-10 microns. It is found on the leaves of tomatoes.

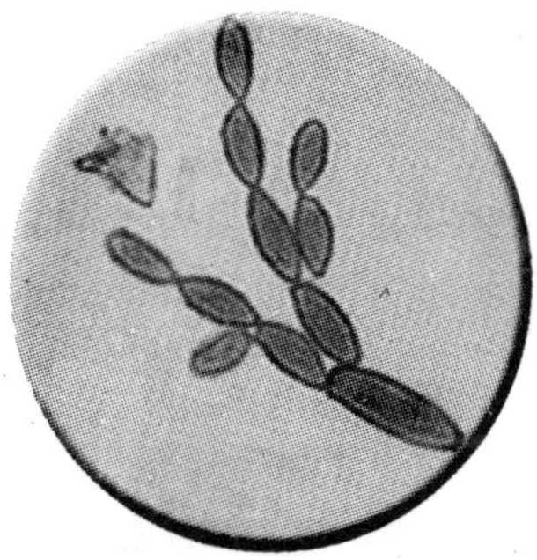

- **cladosporium herbarum**
  conidia dimensions 5-23 x 3-8 microns. It is found on dead plants, woody plants, food, straw, soil, paint and textiles.

- **cladosporium macrocarpum**
  conidia dimensions 9-29 x 5-13 microns. It is found on dead plants, woody plants, food, straw, soil, paint and textiles.

- **cladosporium (hormodendrum)**
most commonly identified outdoor fungus. The outdoor numbers are reduced in the winter. The numbers are often high in the summer. Often found indoors in numbers less than outdoor numbers. It is a common allergen. Indoor Cladosporium sp. may be different than the species identified outdoors. It is commonly found on the surface of fibreglass duct liner in the interior of supply ducts. A wide variety of plants are food sources for this fungus. It is found on dead plants, woody plants, food, straw, soil, paint and textiles. It can cause mycosis. Produces greater than 10 antigens. Antigens in commercial extracts are of variable quality and may degrade within weeks of preparation. Common cause of extrinsic asthma (immediate-type hypersensitivity: type I). Acute symptoms include edema and bronchios-pasms. Chronic cases may develop pulmonary emphysema.

- **cladosporium sphaerospermum**
conidia dimensions 3-4.5 microns. It is found as a secondary invader of plants, food, soil, paint and textiles.

- **clamp connection**
a re-curving outgrowth of a cell that, at cell division, acts as a bridge to allow passage of one of the products of nuclear division into the penultimate cell, thereby assuring maintenance of the dikaryotic condition (of members of the Basidiom-ycotina).

- **classification**
1. arrangement of organisms into groups based on mutual similarity or evolutionary relatedness.
2. systematic arrangement of soils into groups or categories on the basis of their characteristics.

- **clavate**
club-shaped.

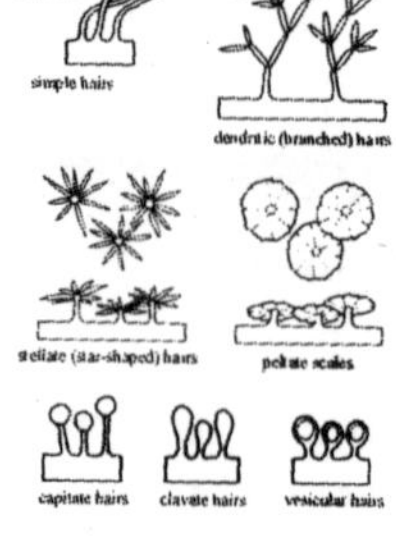

• **clay**
soil particle less than 0.002 mm in diameter.

• **clay aggerate**
a product that is manufactured exclusively in high tech kilns in Germany and used as a soil replacement on hydroponics. It once was very popular in the late '70's-'80's. Now making a comeback.

• **clean seed**
usage: For tall fescue seed, endophyte-free seed is the preferred term.

• **cleistogamous**
With fertilisation taking place within the unopened flower

Cleistogamous flowers in *Viola*

• **cleistothecium**
an ascocarp with the asci surrounded by fungal tissue and without regularly formed openings. (pl. cleistothecia).

• **climax**
most advanced successional community of plants capable of development under, and in dynamic equilibrium with, the prevailing environment.

• **climbers**
those gardeners who are willing to hike for distances to see an alpine specimen.

• **clinandrium**
a cavity, at the apex of a column in orchids, in which the anthers rest.

• **cline**
1. a character gradient within a species, related to geographic distribution.
2. a more or less uniform series of variants from one extreme to the other.

• **clippings**
leaves, and in some cases stems, cut off by mowing.

• **cloaca**
in nematodes: a common duct or cavity in which the digestive and reproductive systems terminate in males.

• **cloche**
a small protective structure made of glass or plastic on a metal or wooden frame-

work. It is placed over delicate plants in the garden to protect them from frost or to warm the soil before planting out seedlings or sowing seeds.

- **clone**

1. a population of recombinant DNA molecules all carrying the same inserted sequence, a colony of microorganisms containing a specific DNA fragment inserted into a vector, a population of cells or organisms of identical genotype.

2. the use of in vitro recombination techniques to insert a particular DNA sequence into a vector, the selection of a unique virus isolate from individual plaques, pocks or lesions or by limiting dilution, the vegetative propagation of an organism to produce a population of identical individuals.

- **cloning**

an in vitro procedure in which a particular sequence of DNA (e.g., a gene) is reproduced in large amounts by inserting ('splicing') it into a suitable replica, introducing the resultant recombinant (hybrid molecule) into a cell in which it can replicate, and finally growing the cells in culture.

- **cloning vector**

DNA molecule that is able to bring about the replication of foreign DNA fragments.

- **closterovirus**

from Greek kloster, 'thread' Member of a group of plant viruses with very long, flexuous, rod-shaped particles containing a single molecule of linear RNA, some members of which are transmitted by whiteflies.

- **club root**

a disease of cabbages and some related vegetables caused by the slime mould fungus.

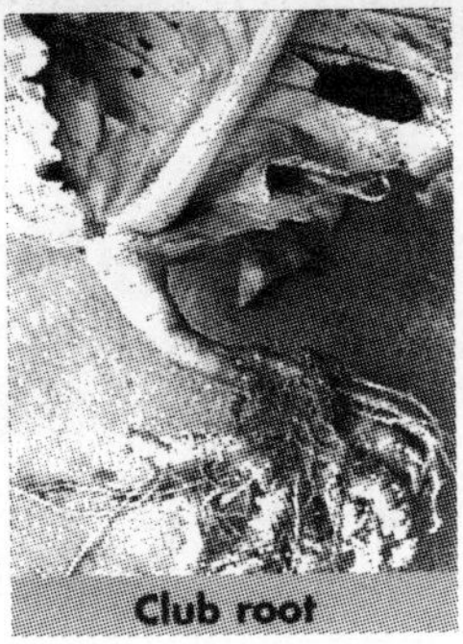

- **cluster cup**
  aecidium.
- **coalesce**
  to merge or grow together into a similar but larger structure.
- **coated soluble fertiliser**
  see **slow-release fertiliser**.
- **coccus**
  a spherical (or near-spherical bacterial cell).
- **codon**
  a particular sequence of three nucleotides in MRNA coding for an amino acid
- **coelomycetes**
  a group of the Deuteromycetes producing pycnidia or acervuli
- **coenocyte**
  a multinucleate cell, a protoplast in which the nuclear divisions have not been followed by cytoplasmic cleavage
- **coenocytic**
  multinucleate or without cross walls. See **syncytium.**
- **coenozygote**
  a cell containing more than one zygote.
- **coenzyme**
  low-molecular-weight chemical which participates in an enzymatic reaction by accepting and donating electrons or functional groups.
- **cohesion**
  force holding a solid or liquid together, owing to attraction between like molecules.

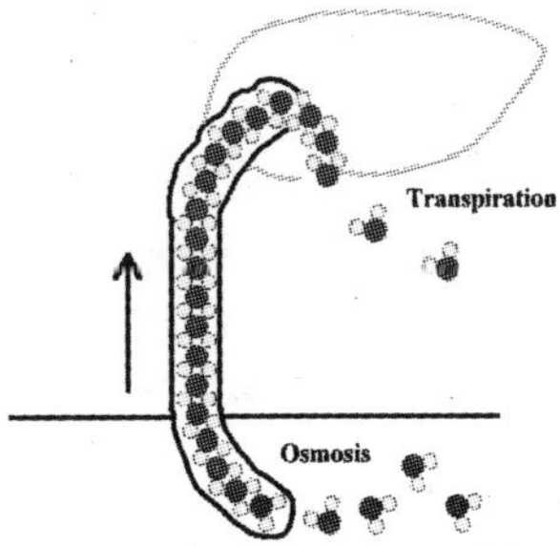

- **cold compost**
  a method by which organic material just rots on its own. It may take months or years to naturally decompose. There may be a significant

amount of weed seeds. And, there may be the danger of some disease organisms still in the compost.

- **cold composting**
  composting under conditions where the temperatures do not rise to 140 °F.

- **cold frame**
  similar to a cloche, this is usually a fixed structure alongside a greenhouse and mostly used to protect plants in pots or seed trays.

- **cold test**
  a seed vigour test in which seeds are planted for a period of time in cool, moist un sterilised soil before transfer to a higher temperature, designed to approximate unfavourable planting conditions.

- **cold-water soluble nitrogen**
  a form of fertiliser nitrogen soluble in cold water (25 °C, 77 °F).

- **cole crops**
  these are members of the cabbage family (i.e. broccoli, cabbage, cauliflower, kale, collards, kohlorabi)

- **coliform**
  gram-negative, no spore-forming facultative rod that ferments lactose with gas formation with in 48 hours at 35°C. Often an indicator organism for faecal contamination of water supplies. *Escherichia coli* and *Enterobacter* are important members.

- **collar**
  a band of material used as a mechanical barrier to protect a plant from damage by insects.

- **colloid fraction**
  organic and inorganic matter with very small particle size and a correspondingly large surface area per unit of mass.

- **colonisation**
  establishment of a community of micro-organisms at a specific site or ecosystem.

- **colony**
  clone of bacterial cells on a solid medium that is visible to the naked eye.

- **colourant**
  a dye, pigment, or paint like material applied to turf to create a favourable green colour when the grass is discoloured or damaged.

- **coloured leaf**
  leaves with one or more colours apart from green, white or cream are distinctly present.

- **column**
  the male and female reproductive organs of the orchid. The column (technically called a 'gynostemium') is formed by the fusion of male portion of the flower (stamens) and female portion (pistils). This is one major characteristic that defines orchids and differentiates them from all other flowering plants.

- **column-foot**
  a basal extension of the column to which the labellum is attached.

- **combing**
  using a comb to lift stolons and procumbent shoots so that they can be cut by a mower the comb typically has metal teeth or flexible tines and is fastened immediately in front of a reel mower.

- **cometabolism**
  transformation of a substrate by a micro-organism without deriving energy, carbon, or nutrients from the substrate. The organism can transform the substrate into intermediate degradation products but fails to multiply at its expense.

- **commensalism**
  interaction between organisms where one organism benefits from the association while the second organism remains unaffected.

- **common name**
  the name by which plants are known by non-bota-

nists. Plants that have a short history of cultivation may not have a common name. These names vary from country to country even from region to region.

- **community**
all organisms that occupy a common habitat and interact with one another.

- **comovirus**
siglum of cowpea mosaic virus. Member of a group of multi-component plant viruses with small, isometric particles containing two linear RNA species, readily transmitted mechanically and by beetles

- **compaction**
often this term comes up when one is talking about new landscaping around a new construction whether it be a private home site, or commercial site. Compaction is created by heavy machinery squeezing the layers of the soil together. It is destructive to the composition and structure of the soil. No longer are there healthy air pockets for roots. The soil is no longer of good texture for planting. Often nutrients are washed away due to poor drainage, or no drainage at all.

- **companion crop**
one crop sown with another, used particularly in the small grains with which forage crops are sown. Synonym nurse crop. Usage: Companion crop is the preferred term.

- **companion planting**
different plants that are planted together for the benefit of each other. Whether it be colour or roots deeper to bring up the nutrients for the secondary plant. Ground covers are great companion plants.

- **compartmentalisation**
in trees: the processes that result in isolation of wounded or diseased xylem from normal xylem by the formation of chemically and

anatomically specialised tissue around the damaged zone.

- **competent**
in a genetic sense, the ability to take up DNA.

- **competition**
1. rivalry between two or more species for a limiting factor in the environment that usually results in reduced growth of participating organisms.
2. a more or less active demand on the part of two organisms for some commodity (space, food, etc). that is inadequate to provide for all organisms present.

- **complementary**
in reference to base pairing, the ability of two poly nucleotide sequences to form a double-stranded helix by hydrogen bonding between bases in the two sequences.

- **complete fertiliser**
a fertiliser that can provide all the three main elements: nitrogen, phosphorus, and potassium.

- **complete metamorphosis**
changes in body form of insects that include egg, larva, pupa and adult, also known as complex metamorphosis.

- **complex medium**
medium whose precise chemical composition is unknown. Also called undefined medium.

- **composite-cross population**
a population generated by hybridising more than two varieties and/or lines of normally self-fertilising plants and propagating successive generations of the segregating population in bulk in specific environments so that natural selection is the principal force acting to produce genetic change. Artificial selection may also be imposed. The resulting population is expected to have a continuously changing genetic makeup. Breeder seed is not maintained as originally released.

- **compost**
usual meaning for the house plant grower is a potting or seed cutting mixture

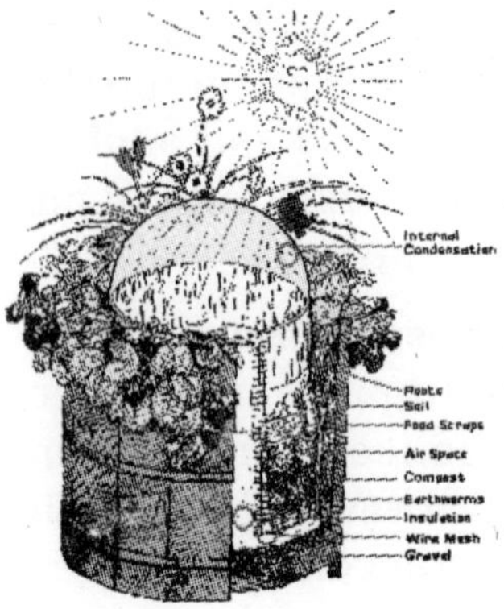

made from peat ('soil less compost') or sterilised soil ('loam compost') plus other materials such as sand, lime, and fertiliser. Compost is also a term for decomposed organic matter such as what is left after a compost heap has degraded vegetable and animal matter.

- **compost heap**
  the result and act of combining organic materials under controlled conditions so that the original raw ingredients are transformed by decay and degradation into humus (or compost).

- **compost tea**
  a low-nutrient liquid that results from placing plant debris in water and allowing it to decompose.

- **composted manure**
  animal faeces that have been aged in a pile, allowing much of the nitrogen to leach from the faeces. A non burning organic fertiliser.

- **compound**
  a structure like a leaf or flower made of one or more parts.

- **compound flower**
  a flower made up of many florets, e.g. Chrysanthemum.

- **compound leaf**
  a leaf made up or two or more leaflets attached to the leaf stalk. E.g. Schefflera.

- **conditioned storage**
  storage of seed under controlled conditions of temperature and relative humidity.

- **conidiogenesis**
  conidium formation.

- **conidiogenous cell**
  a conidium-producing cell.

• **conidioma**
a specialised, multi-hyphal structure bearing conidia (pl. conidiomata.).

• **conidiophore**
a hypha, often specialised in structure, that bears one or more conidia.

• **conidium**
asexual spore resulting from mitotic nuclear division and formed from the ends or sides of a hypha, produced in abundant numbers by the asexual phase of soil fungi in the phyla Ascomycota and Basidiom-ycota. (plural, conidia).

• **conidobolus**
can cause a chronic inflammatory disease of the nasal mucosa (entomophthoromycosis).

• **conifer**
an evergreen, generally green, sometimes cone shaped.

• **conjugation**
in general, any of various sexual processes in micro organisms in which gene transfer follows the establishment of direct contact between two (or more) cells which typically show little or no morphological differentiation from vegetative cells. In bacterial conjugation, one bacterium (the 'male') or donor cell transfers DNA to another (the 'female') or recipient cell while the cells are in physical contact, a recipient that has received DNA from a donor is called a transconjugant.

• **conjugative plasmid**
self-transmissible plasmid, a plasmid that encodes all the functions needed for its own intercellular transmission by conjugation.

• **conk**
the basidiocarp of a wood-decaying fungus, usually a polypore.

• **connation**
connate fusion of like parts. e.g. sepal with sepal: contrasted with adnation.

• **conservatory**
a building build partially or entirely of glass attached to the house and where a large number of plants are grown. Not to be confused with a greenhouse.

• **consortium**
two or more members of a natural assemblage in which each organism benefits from the other. The group may collectively carryout some process that no single member can accomplish on its own.

• **constitutive enzyme**
enzyme always synthesised by the cell regardless of environmental conditions.

• **consumer acceptance**
the likelihood that the consumer will use the genotype, phenotype, or product by choice.

• **contact insecticide**
a poison that must contact the body of the insect to be controlled.

• **context**
the inner or body tissue of a fruit body which supports the hymenophore in the larger and especially the pileate species of Hymenomycetes.

• **continuous stocking**
a method of grazing livestock on a given unit of land where animals have unrestricted and uninterrupted access throughout the time period when grazing is allowed.

• **contractile**
drawing together resulting in decreased size or bulk.

• **control**
economic reduction of crop losses caused by plant diseases.

• **controlled-availability fertiliser**
see **slow-release fertiliser**.

- **conventional planting**
  planting that takes place days to months after a fallow period, usually involves primary and numerous secondary tillage operations. Synonym: delayed planting.

- **convergent improvement**
  a breeding method involving the reciprocal addition to each of two inbred lines of the dominant favourable genes lacking in one line and present in the other. Backcrossing and selection are performed in parallel, each of the original lines serving as the recurrent parent in one series.

- **cool-season crop**
  a crop that grows best during the cool temperatures of spring and fall.

- **cool-season grass**
  turf grasses that actively grow during the cooler spring and fall weather. These include the fescues, rye grass and bent grass.

- **cool-season turf grass**
  turf grass species best adapted to growth during cool, moist periods of the year, commonly having temperature optimums of 15 to 25 °C (60-75 °F).

- **coppicing**
  cutting back trees and shrubs close to the ground to encourage young, vigorous growth, particularly used as a woodland management technique and for the production of willow wands, etc, for fencing and basket weaving. Also see **pollarding**.

- **cordon**
  a very interesting way to grow fruit trees. Apples and pears do well in this mode. The tree is repeatedly pruned and trained to grow as a single rope like stem. For lack of space, this is ideal.

- **core aeration**
increasing air penetration of the soil by removing plugs of soil. A heavy machine with hollow prongs is moved across a lawn pushing the prongs into the soil and pulling out plugs of soil.

- **cored hay samples**
samples taken from stored hay, using a hollow cylinder to remove a core.

- **corm**
a swollen, underground stem base used for propagation, e.g., Crocus.

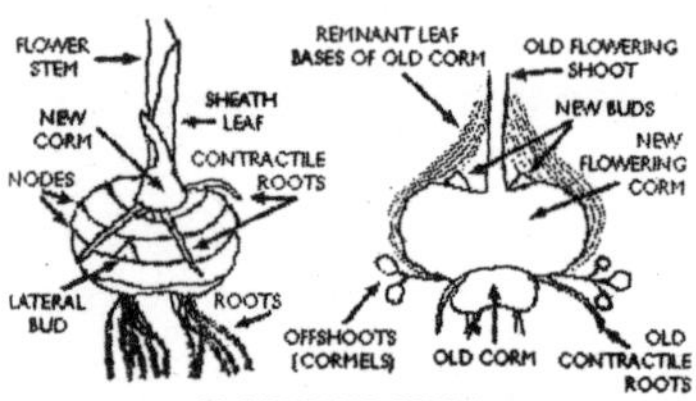

GLADIOLUS CORM

- **cormel**
a small corm that forms around the parent corm. It can be removed and planted to propagate a new plant.

- **corneous**
characterising hard, vitreous, or horny endosperm in cereal grains. See also **chalky**.

- **cornute**
horned, horn-like

- **corolla**
the ring of separate or fused petals which is nearly always responsible for the main floral display.

- **corpus**
in nematodes: The anterior cylindrical part of the esophagus. The basal region of the corpus at times may be swollen to form a bulb.

- **coryneform**
1. essentially rod-shaped with one end thickened or bulbous.
2. a name applied, loosely, to any Gram-positive, asporogenous, pleomorphic rod-shaped bacterium, as such it covers bacteria from a range of genera.

- **cosmid**
a plasmid into which has been inserted the cos site of bacteriophage.

- **cotyledon**
seed-leaf, primary leaf or leaves in an embryo.

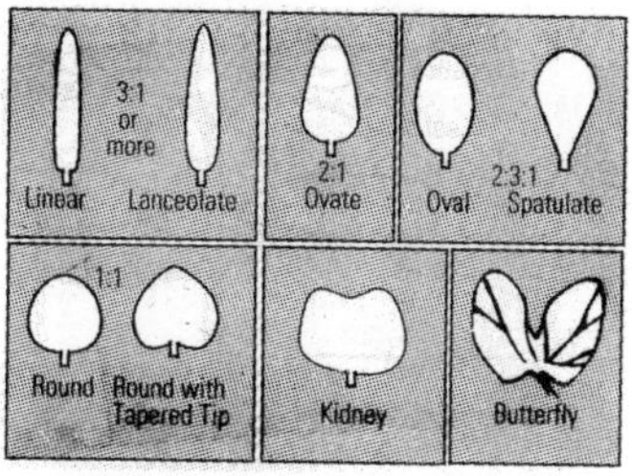

- **coumestrol**
oestrogenic compound occurring naturally in forage crops, especially in ladino clover, strawberry clover, and alfalfa.

- **covalent**
non ionic chemical bond formed by a sharing of electrons between two atoms.

- **cover crop**
a crop that is planted to add humus to the soil or to control weeds (i.e.. winter rye). Usually done between normal planting seasons.

- **cover, winter protection**
see **winter protection cover**.

- **creeper**
any plant that will make long shoots and grow along the ground such as creeping fig, or ivy.

- **creeping growth habit**
plant development by extra vaginal stem growth at or near the soil surface, with lateral spreading by rhizomes and stolons. Synonym: spreading growth habit.

- **crest**
an elevated and irregular or toothed ridge, in orchids found on the lip.

- **crested**
cockscomb-like growth of leaves, stems or flowers. Other name cristate.

- **critical level**
as applied to plant response functions, that level of a factor which is just limiting to plant performance, sometimes specifically defined as 95% of the performance when the factor is non-limiting Usage: Widely used in plant and soil analysis for evidence of nutrient deficiency or toxicity. With leaf-area index, used as the index sufficient for near-maximum light interception or growth rate.

- **crocking**
any material used in the bottom of containers to pro-

vide drainage (i.e., shells, rocks, broken pottery, Styrofoam).

- **crocks**
pieces of broken earthenware pots put in the bottom of other pots to give free drainage and air circulation as well as the prevention of compost or soil blocking the drainage hole.

- **crop residue**
portion of plants remaining after seed harvest, refers mainly to grain crop residue, such as corn stover, or of small-grain straw and stubble.

- **crop rotation**
the practice of growing a sequence of different crops on the same land in successive years or seasons, done to replenish the soil, curb pests, etc.

- **cross**
the progeny resulting from pollination from one plant to another. The term is sometimes applied to a hybrid between different species. 'Cross' is also used to describe transferring of pollen from one flower of a plant to another flower of a different plant.

- **cross fertilisation**
the fertilising of a flower by the transfer of pollen from another flower.

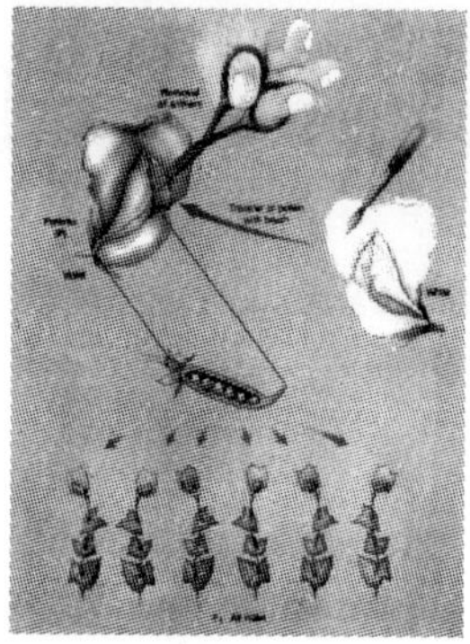

- **cross pollination**
the transfer of pollen from the flower of one plant to the flower on a different plant. Many species require this to set seed.

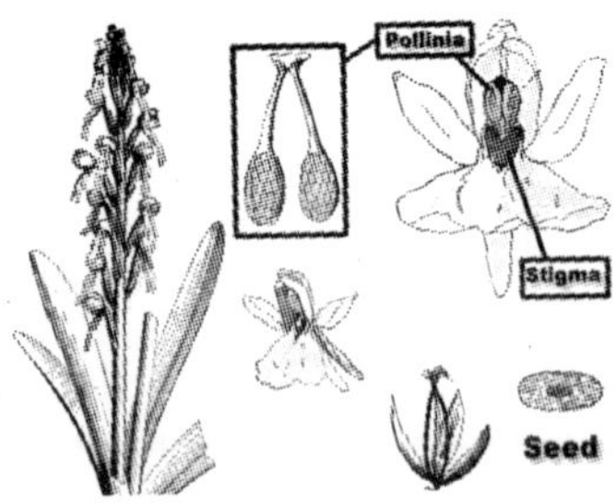

- **cross-feeding**
1. specific type of syntrophy where two populations co-

operate to metabolise a compound.
2. one organism consuming products excreted by another organism.

- **cross-protection**
the protection conferred on a host by infection with one strain of a virus that prevents infection by a closely-related strain.

- **crotch**
the angle measured from the trunk of a tree to the upper surface of a branch.

- **crown**
usually at ground level, this is the part of the plant where the roots meet the stems and from where new shoots grow. The crown also refers to the part of a tree above the bole.

- **crozier**
a re-curved hook at the tip of an ascogenous hypha, the penultimate cell of which will become the ascus.

- **crozier formation**
process of ascus development from coiled tips of ascigerous hyphae.

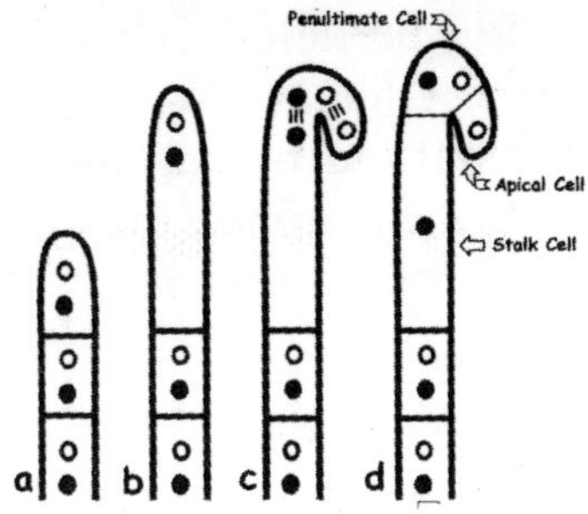

- **crucifer**
any plant in the crucifer of mustard family. Those flowers with four petals are arranged like a cross.

- **crustaformeria**
in nematodes: Glandular region of the distal part of uterus that may play a role in the formation of the egg envelope, sometimes called the quadricolumella.

- **cryptostroma corticale**
conidia dimensions 4-6.5 x 3.5-4 microns. Found on the bark of maple and sycamore trees and on stored logs.

- **cubing**
process of forming hay into high-density cubes to facilitate transportation, storage, and feeding.

- **cucumovirus**
siglum of cucumber mosaic virus. Member of a group of multicomponent plant vi-

ruses with isometric (icosahedral particles) consisting of three linear RNA species, transmitted by sap and in nature by aphids in a non circulative manner.

- **culinary**
  used in cooking.
- **culinary herb**
  a plant grown for its strong flavour which is used to cook with in dishes and salads. the parts of the plant used are the leaves, flowers, or bulbs.

- **cultivar**
  1. a variety, strain, or race that has originated and persisted under cultivation or was specifically developed for the purpose of cultivation.
  2. for cultivated plants, the equivalent of botanical variety, in accordance with the International Code of Nomenclature of Cultivated Plants, 1980.
- **cultivate**
  breaking the topsoil so water and air can penetrate, and, to prevent weeds.
- **cultivation**
  the technique of weeding and hoeing for the purpose of increasing the air in the to layers of the soil and to break up the soil so water will penetrate.
- **cultural control**
  the use of good gardening techniques to control pest populations.
- **culture**
  population of micro organisms cultivated in an artificial growth medium. A pure culture is grown from a single cell, a mixed culture consists of two or more microbial species or strains growing together.
- **culture collection**
  a repository of cultures of characterised viruses, bacteria, and other organisms. Used for reference and comparison with new isolates.
- **culun**
  in the bamboo world this refers to the stem of grasses being usually hollow.

- **cunninghamella**
can cause disseminated and pulmonary infections in immune compromised hosts.

- **cup cutter**
a hollow cylinder with a sharpened lower edge used to cut the hole or cup in a green, or to replace small spots of damaged sod.

- **cupulate**
cup-shaped.

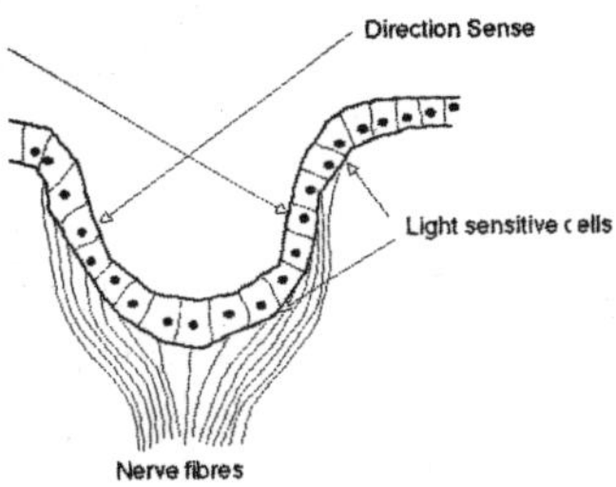

Patella (Limpit)

- **curvularia**
reported to be allergenic. It may cause corneal infections, mycetoma and infections in immune compromised hosts.

- **cut back**
trimming or cutting moderately, making sure some of the last season's growth is left, to clean the plant up and the encourage new growth.

- **cuticle**
1. a thin, waxy layer on the outer wall of epidermal cells consisting primarily of wax and cutin.
2. non-cellular exterior covering of nematodes.

FORMATION OF NEW EXOSKELETON

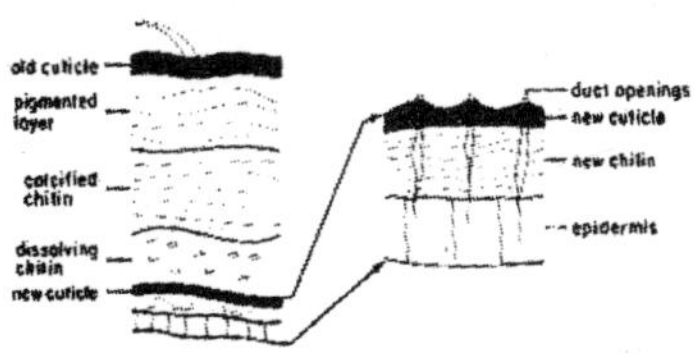

- **cutin**
an insoluble polymer that, embedded in waxes, forms the cuticle covering the epidermal cell walls in the aerial parts of higher plants.

- **cutinolytic**
of certain enzymes: able to digest cutin.

- **cutis**
of basidiocarps of certain wood-decaying fungi: the outer layer consisting of compressed hyphae parallel to the surface, sometimes with varnish like incrustation.

- **cutout**
in cotton: the occurrence of physiologically indeterminate growth.

- **cutting**
  this can be a leaf, roots, shoot, or a bud that has been cut off and then used in propagation.

- **cyanobacterium**
  prokaryotic, oxygenic phototrophic bacterium containing chlorophyll a and phycobilins, formerly the 'blue-green algae.'

- **cyanogenesis**
  the release of hydrocyanic acid (HCN), in the process of chemical change. adj. Cyanogenetic, cyanogenic. Usage: cyanogenetic is the preferred adjectival form.

- **cycad**
  an ancient group of plants that were very abundant in the 'age of dinosaurs' (the Jurasic and Cretaceous periods). There are less than 200 species that survive today and are growing in the warmer regions of the world. Often thought of as long-lived flowerless plants. Most are palm or fern-like.

- **cyclic photophosphorylation**
  formation of ATP when light energy is used to move electrons cyclically through an electron transport chain during photosynthesis.

- **cylindrical**
  of the stipe, spores, etc.: of the same diameter throughout the length.

- **cyme**
  a flat-topped or domed flower head in which the flowers at the centre open first.

- **cyst**
  resting stage formed by some bacteria, nematodes, and protozoa in which the whole cell is surrounded by a protective layer, not the same as endospore.

- **cystidium**
  a sterile cell occurring among basidia and often projecting beyond the hymenium, differing morphologically from the basidium (pl. cystidia).

- **cytochrome**
  iron-containing porphyry ring (e.g., haeme) complexed with proteins which act as electron carriers in an electron-transport chain.

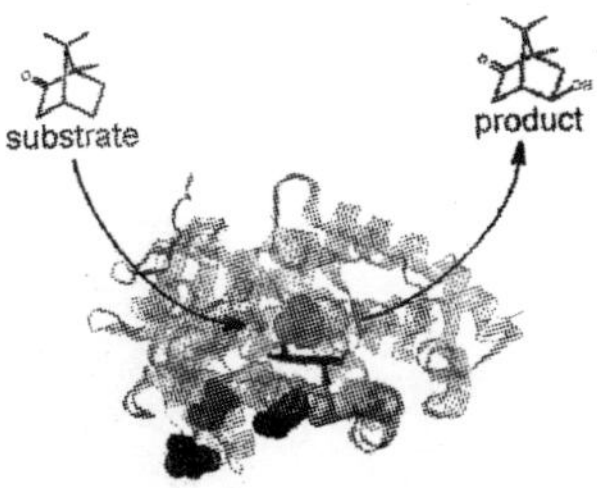

- **cytokinins**
  phytohormones that stimulate metabolism and cell division.

- **cytoplasm**
  cellular contents inside the cell membrane, excluding the nucleus.

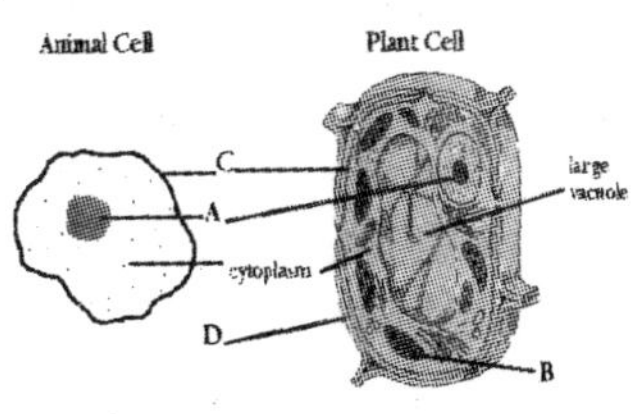

- **cytoplasmic membrane**
  selectively permeable membrane surrounding the cell's cytoplasm.

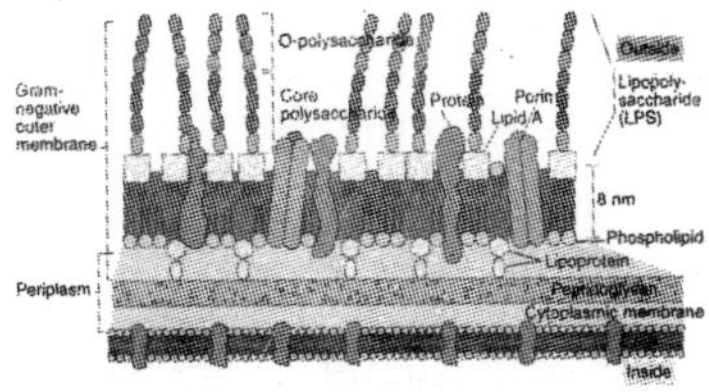

- **damping-off**
  a disease that results in the collapse and death of seedlings before or after they emerge from the soil (pre-emergence and post-emergence damping-off, respectively). Common causal agents include species of Pythium and Rhizoctonia.

- **dappled shade**
  high shade that is created by allowing sun to shine through.

- **dark-dependent seeds**
  seeds that germinate only in darkness. So, must be covered with soil.

- **day length**
  this merely is the number of hours from sunrise to sunset. Sometimes used inter-

changeably with photo period.

- **day neutral**
  a plant whose blooming period is not affected by the length of day.
- **day-neutral plant**
  a plant that will flower under any day length.
- **days to maturity**
  the number of days between planting the seed and first harvest.
- **deadheading**
  1. pinch here, snip there, removing spent flowers that have already bloomed. This is done for the benefit of the plant to prevent disease, prevent seed development and will encourage more vigorous blooming and a bushier plant.
  2. to remove spent blossoms of herbaceous plants.
- **decay**
  the gradual decomposition of dead organic matter.
- **decay cycle**
  the changes that occur as plants grow, die, and break down in the soil.

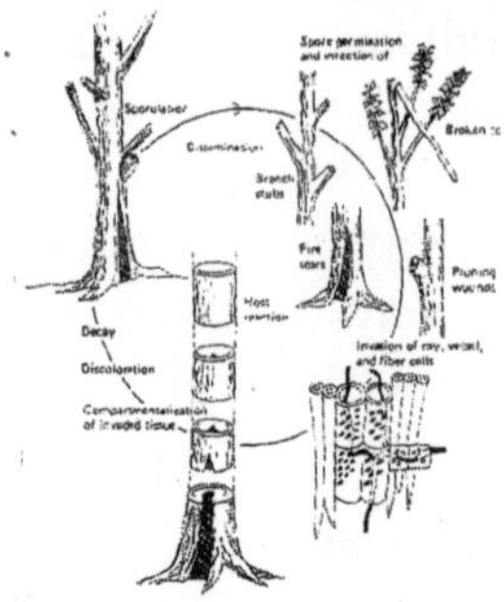

- **deciduous**
  refers to plants that shed their leaves in autumn and then grow them again the following spring. See **evergreen**.

- **decomposer**
  heterotrophic organism that breaks down organic compounds.
- **decomposition**
  chemical breakdown of a compound into simpler compounds, often accomplished by microbial metabolism.

- **deferment**
postponement or delay of grazing on an area, to achieve a specific management objective (e.g., to provide time for plant reproduction, establishment of new plants, restoration of plant vigour, a return to environmental conditions appropriate for grazing, or accumulation of forage for later use).

- **deferred grazing**
the deferment of grazing in a non-systematic rotation with other land units.

- **defined medium**
medium whose exact chemical composition is quantitatively known.

- **degermed corn**
grain (corn) from which the embryo (germ) has been removed.

- **degradation**
process whereby a compound is usually transformed into simpler compounds.

- **dehiscence**
spontaneous opening of a ripe fruit to discharge its seeds.

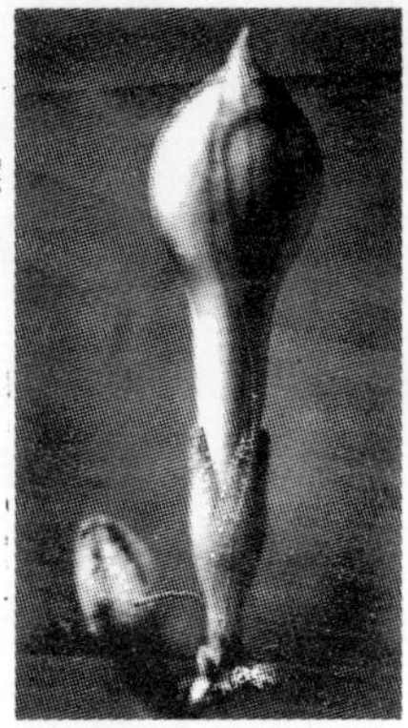

- **dehumidification**
the process by which water is removed from a substance, used specifically of moisture vapour removed from air.

- **dehydration**
an abnormal loss of fluids.

- **delayed-release fertiliser**
see **slow-release fertiliser**.

- **denaturation**
process where double-stranded DNA unwinds and dissociates into two single strands. The reverse of DNA-DNA hybridisation.

- **denitrification**
reduction of nitrate or nitrite to molecular nitrogen or nitrogen oxides by microbial activity (dissimilatory nitrate reduction) or by chemical reactions involv-

ing nitrite (chemical denitrification).

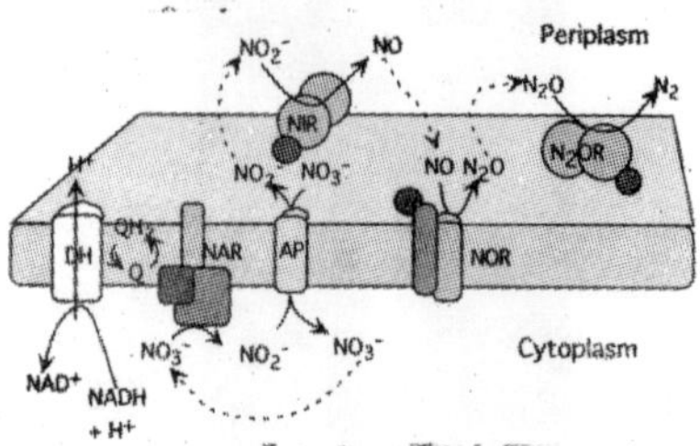

- **deoxyribonucleic acid (DNA)**
  polymer of nucleotides connected via a phosphate-deoxyribose sugar backbone, the genetic material of the cell.

- **derepressible enzyme**
  enzyme that is produced in the absence of a specific inhibitory compound acting at the transcriptional level.

- **dermatophyte**
  a parasitic fungus that attacks and causes a disease of the skin.

- **desert**
  land on which the vegetation is absent or sparse, is usually shrubby, and is characterised by an arid climate, hot to cool.

- **desiccate**
  to dry out.

- **detassel**
  to remove, by cutting or pulling, the tassels that bear staminate flowers in order to prevent selfing during hybrid seed production of corn.

- **determinate**
  having a fixed, definite limit.

- **dethatch**
  the procedure of removing an excessive thatch accumulation either mechanically, as by vertical mowing, or biologically, as by topdressing with soil.

- **deuteromycetes**
  see **deuteromycotina**.

- **deuteromycotina**
  also known as fungi imperfecti. A non-phylogenetic category originally created for fungi with no known sexual stage, the category still includes fungi with no known sexual stage, but it also includes the asexual stages of various fungi now known to have a sexual stage in the Ascomycotina or Basidiomycotina.

• **developmental stage**
discrete portion of the life cycle of a plant, such as vegetative growth, reproduction, or senescence. Several published systems are in use for various crops to subdivide the broad stages. See also bloom, early bloom, full bloom, late boot stage, grain maturity.

• **dextrinoid**
of spores, etc. stained yellowish- or reddish-brown by Melzer's iodine.

• **diagnostic**
a distinguishing characteristic important in the identification of a disease or other disorder.

• **diallelic**
see **balanced diallelic**.

• **diameter breast high**
the diameter of a tree trunk at a height of 4$^1/_2$ feet above the ground.

• **diandrous**
with two stamens, as members of the orchid sub-tribe Cypripedilianea.

• **diatom**
algae with siliceous cell walls that persist as a skeleton after death. Any of the microscopic, unicellular or colonial algae constituting the class Bacillariophyceae.

• **diatomaceous earth**
geologic deposit of fine, greyish siliceous material composed chiefly or wholly of the remains of diatoms. It may occur as a powder or as a porous, rigid material.

• **diazotroph**
organism that can use dinitrogen as its sole nitrogen source, i.e. capable of N_2 fixation.

• **dibber**
a tool for making holes in the soil for individual seeds, bulbs or plants. It is often made from the top 6 to 9 inch of the handle of a broken garden fork or spade giving a useful handle and

then the other end is rounded off.

- **dibble**
  a tool used to make holes for seeds or bulbs: a pencil end, dowel, or anything that works for the situation.
- **dichotomous**
  dividing into two equal branches

- **dicot**
  also dicotyledon, flowering plants with embryos that have two cotyledons.
- **dictyospore**
  a spore divided by intersecting septa in more than one plane.
- **dictyosporous**
  having spores with cross and longitudinal walls.
- **didelphic**
  of nematodes: Possessing two complete genital tubes or ovaries.
- **didymospore**
  a spore with one transverse septum.
- **dieback**
  tips and shoots dying on live plants. Often the dieback will progress up the stem or branch to the main stem so it is important to cut off the dead part.
- **differential host**
  a plant host that on the basis of disease symptoms serves to distinguish between various strains or races of a given plant pathogen.
- **differential medium**
  cultural medium with an indicator, such as a dye, which allows various chemical reactions to be distinguished during growth.

- **differentiation**
  the process leading to differences in the amount, character, and location of cellular components, cells, tissues, and organs of an organism: specialisation physically, chemically, physiologically, and morphologically. In particular, the maturation of a cell with unique traits (e.g., a xylem vessel), different from the meristematic cells from which it arose and different from other cells of similar age that followed a different course.

- **diffusion (nutrient)**
  movement of nutrients in soil that results from a concentration gradient.

- **digestibility, dry matter**
  see **in vitro dry matter disappearance**.

- **digestible dry matter**
  feed intake minus faeces expressed as a percent of feed dry matter consumed.

- **digestible energy intake**
  feed consumption expressed as units of digestible energy.

- **digestible protein**
  feed protein minus faeces portion (nitrogen x 6.25) expressed as a percentage of amount in feed.

- **dikaryon**
  two nuclei present in the same hyphal compartment, they constitute a homokaryon when both

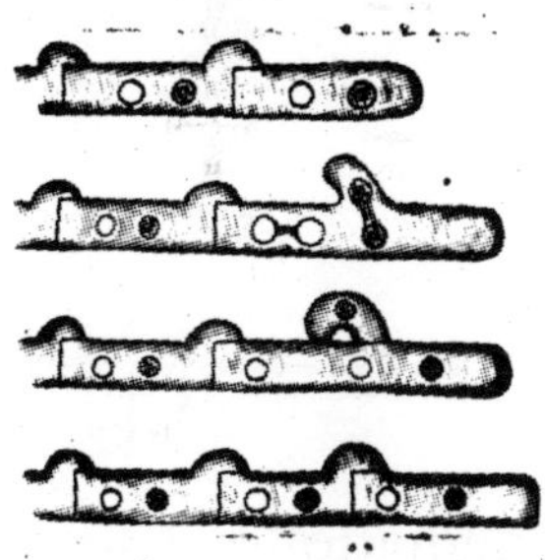

nuclei are genetically the same or a heterokaryon when each nucleus is genetically different from the other.

- **dikaryotic**
  the condition of containing a dikaryon.

- **dilution plate count method**
  method for estimating the viable numbers of microorganisms in a sample. The sample is diluted serially and then transferred to agar plates to permit growth and quantification of colony-forming units.

- **dimorphic**
  producing two morphologically different forms.

- **dinitrogen fixation**
conversion of molecular dinitrogen (N_2) to ammonia and subsequently to organic

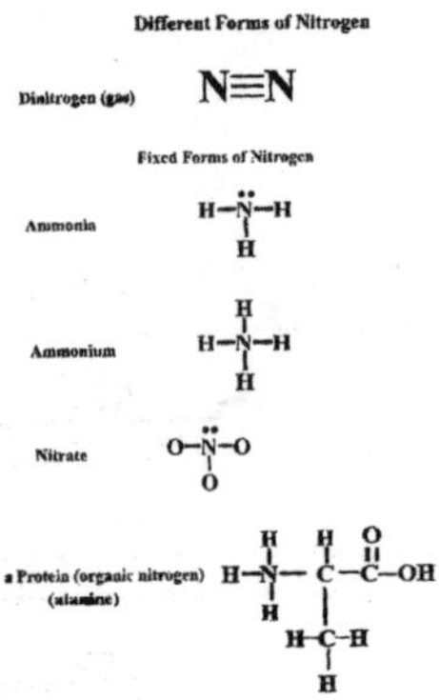

combinations or to forms useful in biological processes.

- **diocecious**
a plant which bears either male or female flowers.

- **diploid**
in eukaryotes, an organism or cell with two chromosome complements, one derived from each haploid gamete.

- **direct count**
method of estimating the total number of micro organisms in a given mass of soil by direct microscopic examination.

- **direct seeding**
putting the seed directly in the soil as opposed to transplanting seedlings.

- **dirty seed**
usage: For tall fescue seed, endophyte-infected seed is the preferred term.

- **disbudding**
in reference to fruit crops, by selectively taking off buds to diminish the crop production and to have quality over quantity.

- **disc**
of Valsa and related fungi: a more or less flat apical part of a stroma that protrudes above the bark surface, also, of Discomycetes: the exposed fertile portion of an apothecium.

- **discomycetes**
a group of the Ascomycetes in which the hymenium is exposed at maturity, one in which the fruiting body is an apothecium or discocarp.

- **disease**
  an abnormal condition of a plant in which its physiology, morphology, and development is altered under the continuous influence of a pathogen.

- **disease cycle**
  of a disease caused by a biotic agent: the cyclical sequence of host and parasite development and interaction that result in disease, in reproduction or replication of the pathogen, and in the readying of a new generation of the parasite for infection.

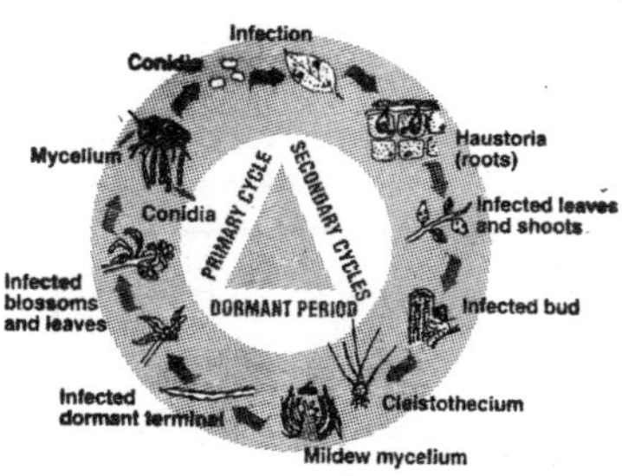

- **disease resistance**
  the tendency not to be infected by a particular pathogen.

- **disease tolerance**
  the ability of a plant to continue growing without severe symptoms despite being infected by a pathogen.

- **dish garden**
  this arrangement is most seen in florists. Many plants grown together to be used indoors for a focal point of greenery.

- **disinfectant**
  a physical or chemical agent that frees a plant, organ, or tissue from infection.

- **disinfest**
  to kill pathogens that have not yet initiated disease, but that occur in or on such inanimate objects as soil, tools, etc., or that occur on the surface of such plant parts as seed.

- **dispersal**
  spread of a pathogen within an area of its graphical range.

- **dissemination**
  see **dispersal**.

- **DNRA**

  Dissimilatory Nitrate Reduction to Ammonium. The use of nitrate by organisms as an alternate electron acceptor in the absence of oxygen resulting in the reduction of nitrate to ammonium.

- **distichous**

  in two ranks or rows on opposite sides of an axis.

- **distilled water**

  pure water free from dissolved salts. Formerly made by distillation, now produced chemically by demineralisation.

- **dither**

  just a plain old utensil of any kind to make a hole in the ground to drop a seed into.

- **diurnal**

  referring to daytime, in reference to flowers, signifying those which open only during the day.

- **division**

  a method of propagating plants by separating each one into two or more sections and then repotting (i.e. perennials are easily propagated this way).

- **DNA**

  Deoxyribonucleic Acid.

- **DNA fingerprinting**

  a laboratory technique in which the banding patterns of DNA fragments from two different individuals are compared.

- **DNA library**

  collection of cloned DNA fragments which in total contain genes from the entire genome of an organism, also called a gene library.

- **DNA polymerase**

  an enzyme that synthesises a daughter strand(s) of DNA (under direction from a DNA template). May be involved in repair or replication.

JOINING OF OKAZAKI FRAGMENTS

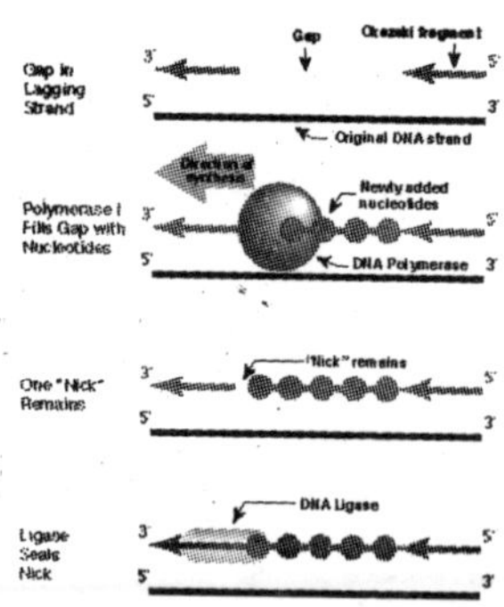

- **DNA replicase**

  a DNA-synthesising enzyme required specifically for replication.

• **DNAASE**
an enzyme that attacks bonds in DNA.

• **dolipore septum**
specialised crosswall separating compartments of a hypha of fungi in the phylum Basidiomycota, consisting of a central pore covered with perforated membranes on both sides (called a parenthosome).

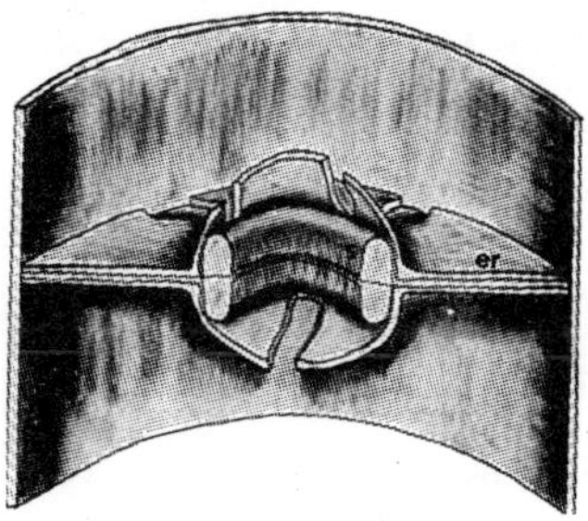

• **dolomitic limestone**
sometimes used when 'liming' soil that has an acid pH level that is too high. As it contains calcium and magnesium carbonate it should be used only with soils that are also deficient in magnesium as well.

• **domain**
highest level of biological classification, superseding kingdoms. The three domains of biological organisms are the Bacteria, the Archaea, and the Eukarya.

• **dormancy**
a state of suspended growth or lack of visible activity caused by environmental or internal factors.

• **dormant oil**
a great horticultural oil to be used on fruit trees or any plant material that has insects. It mainly kills the eggs that are not seen. Read the directions carefully not to be used in high temperatures. There are several brands on the market.

• **dormant period (dormancy)**
this is the time when a plant has naturally stopped growing and the leaves have fallen or the top growth has died down. The dormant period is usually, but not always, in winter. Most plants need it to perform.

• **dormant seeding**
planting seed during late fall or early winter after temperatures become too low for seed germination to occur until the following spring.

- **dormant turf**
turfs that have temporarily ceased shoot growth as a result of extended drought, heat, or cold stress.

- **dorsal**
back or upper surface.

- **double dig**
a method of digging a garden bed which involves removing the soil to the depth of one spade blade and then digging down an equal distance, breaking up and mixing the soil.

- **double flower**
the Latin name for this is 'flore pleno.' It refers to flowers that have many petals present, such as roses.

- **double potting**
placing a potted plant in a larger pot with damp peat moss surrounding it. The peat is kept moist and provides a humid evaporative effect for the potted plant nestled between it. Used a lot to dress up a working clay pot.

- **doubling time**
time needed for a population to double in number or biomass.

- **downy mildew**
a certain kind of mildew caused by a special fungi. Often confused with sooty mildew and powdery mildew. As with all mildews it is a problem in hot and humid weather. And, like many fungi it transports its spores. The plants affected will have fuzzy patches on the leaves.

- **drainage**
how water moves through the soil. A real important factor for most plants and gardens. In general water should move through the soil whether in a garden or in a container somewhat easily. If there is standing water create better drainage by adding non-porous material.

- **drawn**
referring to the structure of a plant, one that is too tall and has grown too weak.

Caused by growing in too little light or too close together. Often is the case in flats in nurseries.

- **dreschlera**

  conidia dimensions 40-120 x 17-28 microns. Found on grasses, grains and decaying food. It can occasionally cause a corneal infection of the eye.

- **drift**

  a design term generally attributed to Gertrude Jekyll. To express a feeling in with plants. The technique is to plant flowers thicker in the centre and further apart on the outskirts.

- **drill**

  a shallow straight furrow for sowing seeds.

- **drip irrigation**

  a system of tubes with small holes that allow water to drip out onto the root zone of plants. A water-conserving irrigation system.

- **drip line**

  the imaginary line under the tips of the canopy of a tree.

- **drought**

  a prolonged period of dryness that can cause damage to plants.

- **drupe**

  a type of fruit (e.g., plums, cherries, olives, peaches). Also considered stone fruits. The fruit wall is fleshy. The outside layer is generally juicy. The one seed fruit will not open up.

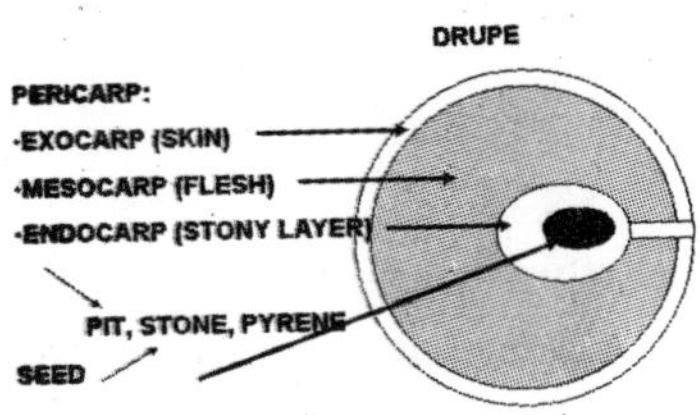

- **dry matter**

  the substance in a plant remaining after oven drying to a constant weight at a temperature slightly above the boiling point of water.

- **dry matter disappearance**

  1. grazing: forage present at the beginning of a grazing period plus growth during the period minus forage present at the end of the period.
  2. digestibility: loss in dry weight of forage exposed to in vitro digestion.

- **dry milling**
  see **milling, dry**.
- **dry weight**
  moisture-free weight.
- **dwarf**
  shorter than its normal growth. Each family of plants has a height recommendation for dwarfness.
- **earthing up**
  drawing soil up around the stems of a plant, e.g. potatoes, to blanch them or encourage root or tuber growth.
- **earthworm casts**
  see **castings, earthworm**.
- **ecalcarate**
  without a spur or spurs.

- **echinate**
  furnished with prickles or bristles.
- **echinulate**
  covered with small spines.
- **ecocline**
  series of biotypes within a species that shows a genetic gradient correlated with a gradual environmental gradient.
- **ecological niche**
  role of an organism in an ecological system.
- **ecological race**
  group of local populations within a species in which individuals have similar environmental tolerances. Wide-ranging species may consist of many ecological races.
- **ecology**
  the study of communities of living things and the relationships between organisms and their environment.
- **ecosystem**
  a system consisting of a community of animals, plants and micro-organisms and the physical and chemical environment in which they interrelate.

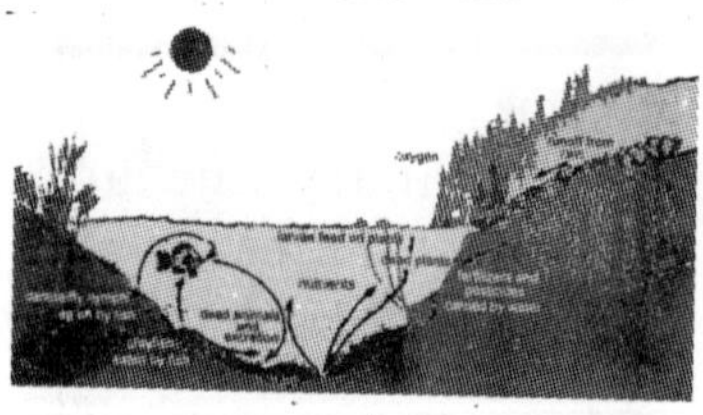

- **ecotone**
  transition zone between two vegetational types or vegetational regions.

- **ectomycorrhiza**
  mycorrhizal type in which the fungal mycelia extend inward, between root cortical cells, to form a network (Hartig net) and outward into the surrounding soil. Usually the fungal hyphae also form a mantle on the surface of the root.

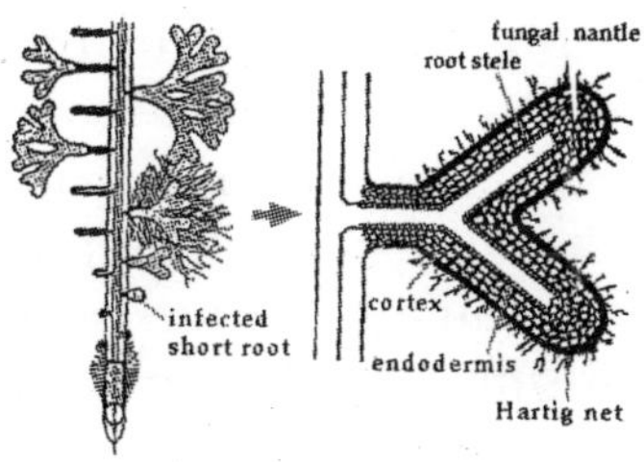

- **ectoparasite**
  a parasite that remains external to the host's cells or tissues

- **ectosymbiosis**
  symbiosis in which one member (micro-symbiote) develops on the outside of the other member.

- **ectotrophic**
  refers to a mycorrhiza in which the mycelium forms an external covering on the root.

- **edaphic**
  1. of or pertaining to the soil.
  2. resulting from or influenced by factors inherent in the soil or other substrate, rather than by climatic factors.

- **edema**
  see **oedema**.

- **edging plant**
  on the edge or border of a bed.

- **eelworm**
  nematode

- **effective dissemination**
  synonymouswith innoculation

- **efficiency**
  degree to which a plant or vegetation converts radiant energy into organic compounds. Efficiency may depend on degree of utilisation of necessary environmental components.

- **efflorescence**
  the deposit of calcium and fertiliser salts on the outer surfaces of clay pots.

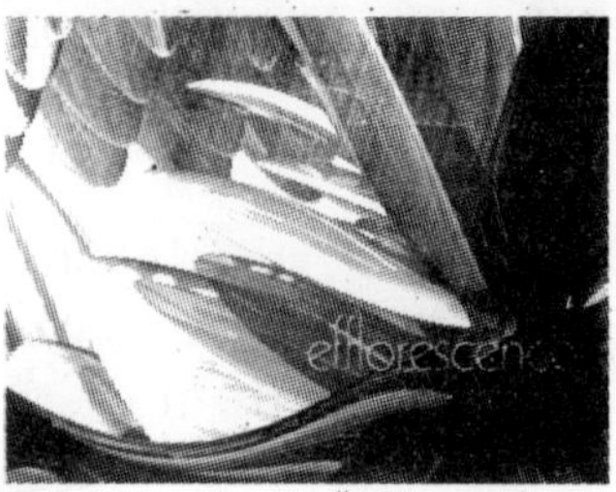

- **effuse**
  spreading out loosely or flat

- **EH**
  potential generated between an oxidation or reduction half-reaction and the H electrode in the standard state.

- **electron acceptor**
  substance that accepts electrons during an oxidation-reduction reaction. An electron acceptor is an oxidant.

- **electron donor**
  substance that donates electrons in an oxidation-reduction reaction. An electron donor is a reductant.

- **electron-transport chain**
  final sequence of reactions in biological oxidations composed of a series of oxidising agents arranged in order of increasing strength and terminating in oxygen.

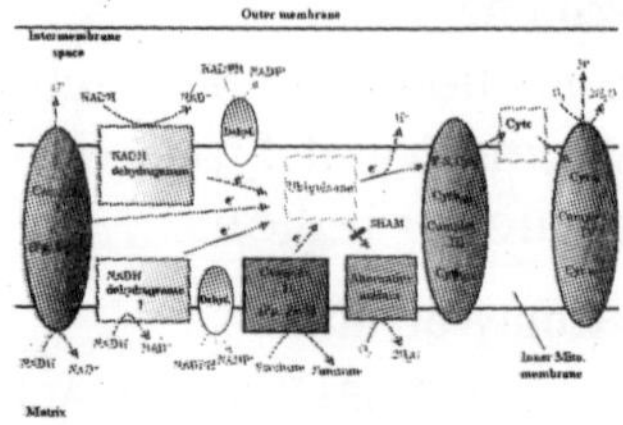

- **electrophilic compounds**
  chemicals that attack or are drawn to regions in other chemicals in which electrons are readily available, oxidising agents act as electrophilic compounds.

- **electrophoresis**
  a procedure by means of which the members of a heterogeneous population of charged particles can be separated by virtue of their dissimilar migration characteristics in an electric field.

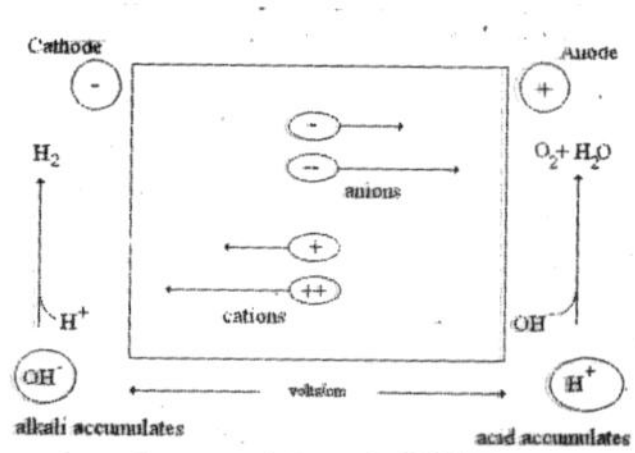

- **electroporation**
  a method by which nucleic acids or virus particles can be introduced into protoplasts or cells by creating transient pores in the

plasma membrane using an electric pulse.

- **element**
  a substance that cannot be separated into different substances. All matter is made of elements.

- **elicitor**
  a molecule produced by the host (or pathogen) that induces a response by the pathogen (or host).

- **ellipsoid**
  having every plane section an ellipse or a circle.

- **ellipsoidal**
  see **ellipsoid**.

- **eluviation**
  removal of soil material from a layer of soil as a suspension.

- **emarginate**
  of gills, notched near the stipe.

- **emasculation**
  the removal of anthers to stop self pollination.

- **embden-meyerhof-parnas pathway**
  a biochemical pathway that degrades glucose to pyruvate, the six-carbon stage converts glucose to fructose-1, 6-bisphosphate, and the three-carbon stage produces ATP while changing glyceraldehyde-3-phosphate to pyruvate.

- **embryo percent**
  1. the amount of embryo compared with endosperm and other seed parts.
  2. the percent of embryo in the whole seed.

- **enation**
  a symptom caused by certain plant viruses in which there are small outgrowths on the plant.

- **enclosure**
  1. an area enclosed by fence or wall to confine animals.
  2. a caged or fenced area within a pasture to exclude grazing.

- **encysted**
  surrounded by a hard shell.

- **endarch**
  primary xylem in which the development is from the centre towards the outside.

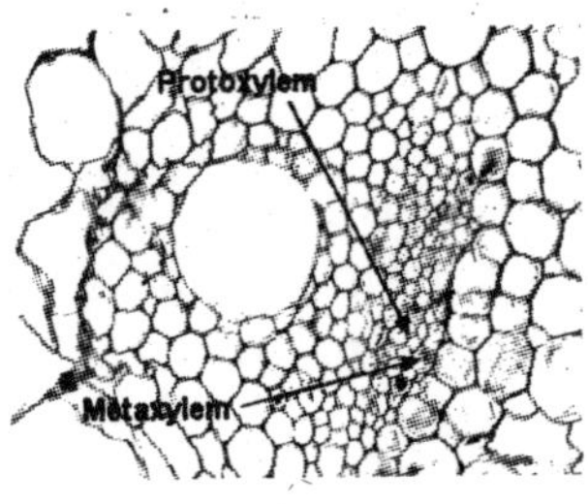

- **endemic**
  plants which are of a certain geographic area and generally are confined to that place.

- **endergonic reaction**
  chemical reaction that proceeds with the consumption of energy.

- **endoconidiophore**
  a conidiophore that produces conidia within itself.

- **endoconidium**
  a conidium produced endogenously in a hypha or conidiophore (pl. endoconidia).

- **endoenzyme**
  enzyme that operates along the internal portions of a polymer.

- **endogen**
  a plant in whose primary stem the vascular bundles are not arranged in concentric circles and whose older stems do not have distinct wood and bark layers.

- **endogenous**
  arising from within the generating structure.

- **endomycorrhiza**
  mycorrhizal association with intracellular penetration of the host root cortical cells by the fungus as well as outward extension into the surrounding soil.

- **endonuclease**
  a nuclease which cleaves phosphodiester bonds within a nucleic acid strand.

- **endoparasite**
  a parasite that lives intracellular or within the tissues of the host.

- **endophyte**
  a plant living within another plant. In turf grasses, it is a fungus within the grass plant secreting substances that repel insect pests.

- **endophyte absence**
  absence of the endophytic fungus *acremonium*

coenophialum in the fescue tissue, plant, or population being discussed.

- **endophyte-free seed**
tall fescue seed free of the *acremonium coenophialum* endophyte.

- **endospore**
differentiated cell formed within the cells of certain Gram-positive bacteria and extremely resistant to heat and other harmful agents.

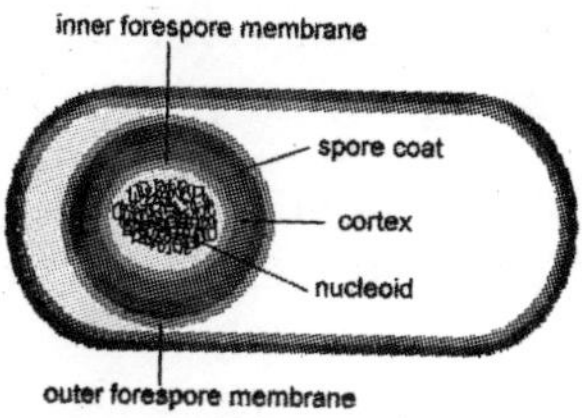

- **endosymbiosis**
symbiosis in which one member (micro-symbiote) lives within the other.

- **endotrophic**
refers to a mycorrhiza in which the mycelium grows within the cortical cells of the root (e.g., in orchids).

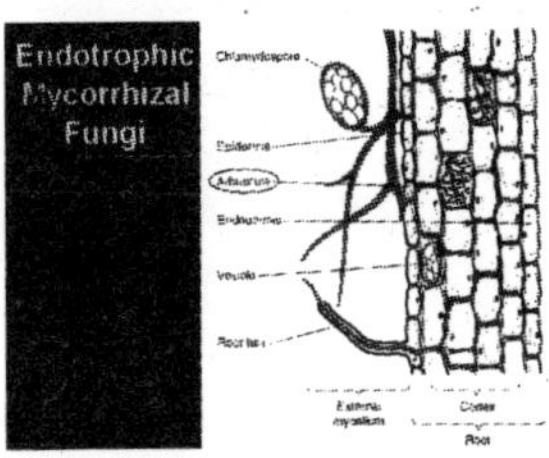

- **enrichment culture**
technique in which environmental (including nutritional) conditions are controlled to favour the development of a specific organism or group of organisms.

- **ensilage**
see **silage**.

- **enteric bacteria**
general term for a group of bacteria that inhabit the intestinal tract of humans and other animals. Among this group are pathogenic bacteria such as Salmonella and Shigella.

- **entire leaf**
an undivided and underrated leaf.

- **enzymatic degradation**
chemical breakdown of a given substance by the specific enzyme catalyst for that particular chemical reaction of a biological process.

- **enzyme**
protein within or derived from a living organism that functions as a catalyst to promote specific reactions.

- **elisa**
enzyme-linked immunosorbent assay. Immu-

munoassay that uses specific antibodies to detect antigens or antibodies. The antibody-containing complexes are visualised through an enzyme coupled to the antibody. Addition of substrate to the enzyme-antibody-antigen complex results in a coloured product.

- **ephemeral**
  lasting only one day when in flower.

- **epichil**
  the upper part of the jointed, complex lip of certain orchids, as in the genus Stanhopea.

- **epichile**
  terminal lobe of labellum in certain orchids.

- **epicoccum**
  conidia dimensions 15-25 microns. It is found in plants, soil, grains, textiles and paper products.

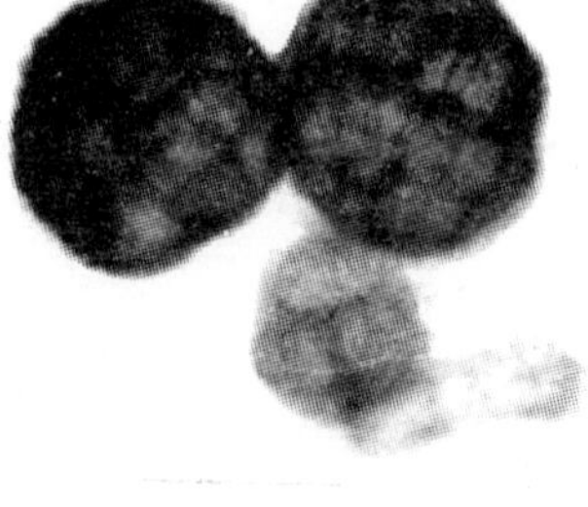

- **epidemic**
  a change in the amount of disease in a population in time and space.

- **epidemiology**
  1. the study of the interrelationships between a given pathogen, the environment, and groups or populations of the relevant hosts.
  2. the study of epidemics.

- **epidermis**
  the superficial layer of cells occurring on all plant parts.

- **epidermophyton**
  can cause infections of skin and nails.

- **epigeal germination**
  a type of germination in which the cotyledons are raised above the ground by elongation of the hypocotyl.

- **epinasty**
  downward curling of a leaf blade resulting from more rapid cell growth on the upper side of a petiole than on the lower side, often a hyperplastic symptom of plant disease.

- **epiparasite**
  an organism parasitic on another that parasitises a third.

- **epiphyte**
  a plant that grows on another plant, such as on a bush or tree, but is not nourished by it (hence, not parasitic). They use the host only for anchorage, drawing food and moisture from the air and from humus collected in the angles of branches or in the crevices of the bark. An 'air-plant.' Orchids generally are found growing one of three ways: as epiphytes (the majority grow in this manner), lithophytes, or terrestrials.

- **epiphytotic**
  see **epidemic**.

- **epiptygma**
  in nematodes: A vulval flap.

- **episome**
  plasmid that replicates by inserting itself into the bacterial chromosome.

- **epitope**
  the region of an antigen to which the variable region of an antibody binds.

- **epitype**
  a specimen selected as a standard for a species or lower taxon when all original material except for illustrations has been destroyed.

- **eradicant**
  any chemical agent that eliminates particular pathogens from diseased plants treated with that agent.

- **eradication**
  control of plant disease by eliminating the pathogen after it is established or by eliminating the plants that carry the pathogen.

- **ergopeptine, ergopeptide**
  a class of alkaloids produced by clavicepitaceous fungal organisms, including ergovaline, ergotamine, and related ergotlike alkaloids.

- **ergot**
  disease of certain grasses and cereals, especially rye,

caused by *claviceps purpurea*, also the spur-shaped sclerotium that replaces the grain in a diseased inflorescence.

- **ericaceous**

either refers to members of the Erica family or to acid loving plants or compost especially produced for growing them in pots.

- **ericoid mycorrhiza**

type of mycorrhiza found on plants in the Ericales. The hyphae in the root are able to penetrate cortical cells (endomycorrhizal habit), however, no arbuscules are formed. Major forms are ericoid, arbutoid, and monotropoid.

- **erose**

with the margin irregularly notched, as if gnawed.

- **erosion**

the wearing away of soil by man, rain, or wind. Not a healthy situation.

- **erumpent**

breaking through the surface, bursting forth.

- **escape**

1. a plant that is on its way to becoming naturalised in an area. Just exactly as it reads, it has escaped from cultivation.
2. failure of inherently susceptible plants to become diseased, even though disease is prevalent.

- **espalier**

a system for growing plants, usually fruit trees, where a plant has one single central stem or trunk with tiers of branches on either side trained to present a flattened aspect. Often used for growing trees against a wall.

- **established**
  the state of a plant when it is adjusted to the site and thriving.

- **ether extract**
  fats, waxes, oils, and similar plant components that are extracted with warm ether in chemical analysis.

- **ethidium bromide**
  a trypanocidal, bacteriostatic dye that binds to DNA and fluoresces under near-ultraviolet light, used for tracking nucleic acids.

- **etiolation**
  a phenomenon exhibited by plants grown in the dark: etiolated plants are pale yellow and have long internodes and small leaves.

- **etiology**
  the study of cause, that phase of plant pathology dealing with the causal agent and its relations with the susceptible plant.

- **eubacteria**
  old term for the Bacteria.

- **eukarya**
  phylogenetic domain containing all eukaryotic organisms.

- **eukaryote**
  organism having a unit membrane-bound nucleus and usually other organelles.

- **eutrophic**
  having high concentrations of nutrients optimal, or nearly so, for plant or animal growth. Can be applied to nutrient or soil solutions and bodies of water.

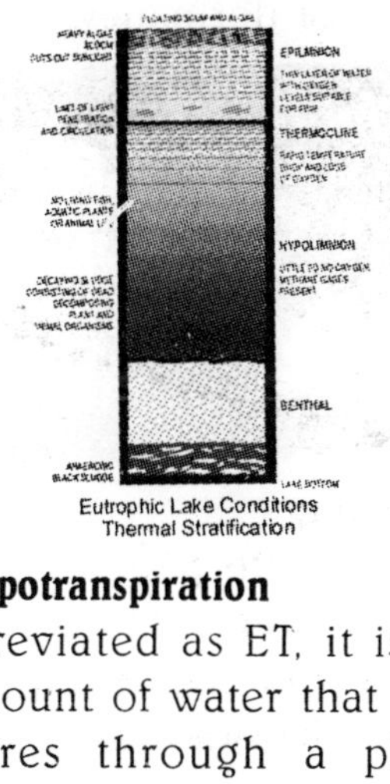

Eutrophic Lake Conditions
Thermal Stratification

- **evapotranspiration**
  abreviated as ET, it is the amount of water that transpires through a plants leaves combined with the

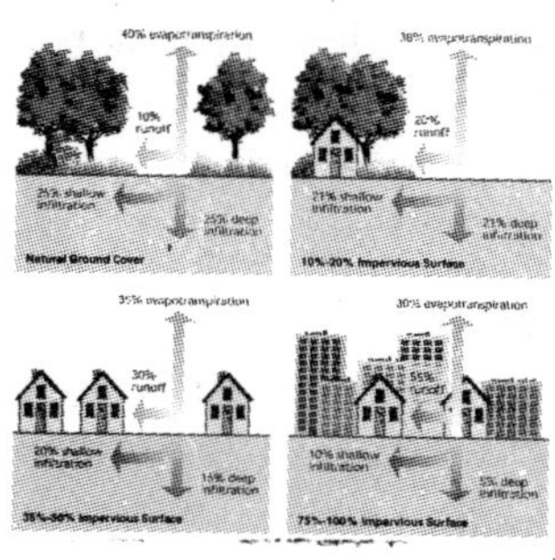

amount that evaporates from the soil in which it is growing. Used as a guide for how much water a plant needs per day.

- **everblooming**
  those flowers that will bloom all season.

- **evergreen**
  the opposite of deciduous, a plant that does not shed leaves in winter or other period of dormancy.

- **everlasting**
  flowers with papery petals which retain some or all of their colour when dried for winter decorations.

- **exarch**
  primary xylem in which the development is from the outside, toward the centre. Typical of roots.

- **exclusion**
  the principle of plant disease prevention in which the pathogen is prevented from entering a given region.

- **exergonic reaction**
  chemical reaction that proceeds with the liberation of energy.

- **exobiology**
  branch of biology concerned with the effects of extraterrestrial environments on living organisms.

- **exoenzyme**
  enzyme that acts at the end of a polymer cleaving off monomers and dimmers and sometimes larger chain fragments.

- **exogenous**
  arising on the outside of the generating structure.

- **exonuclease**
  a nuclease that sequentially removes nucleotides from one end of a strand of nucleic acid.

- **exopathogen**
  nonparasitic organism whose extra-cellular toxic metabolites cause disease in plants.

- **exotic**
  strictly speaking, a plant which is not native to the

area, but popularly any unusual or striking plant.

- **explant tissue**
  the excised tissue of a plant (or seed) used to establish a cell or tissue culture system.

- **exponential growth**
  period of sustained growth of a micro-organism in which the cell number constantly doubles within a fixed time period.

- **exponential phase**
  period during the growth cycle of a population in which growth increases at an exponential rate. Referred to as logarithmic phase.

- **extra-cellular**
  outside the cell.

- **extrusion cooking**
  a process where a mash or granular mixture of starch and/or proteins of cereals or oilseeds is pre-conditioned by exposure to steam and then cooked in a continuous extruder.

- **exudate**
  material that has passed from within a plant structure to the outer surface or into the surrounding medium, as in leaf exudate, root exudate, etc.

- **eye**
  two unrelated meanings: an undeveloped growth bud (as in a potato) or the centre of a flower.

- **F1**
  breeders use this term and it refers to the first generation offspring, from two plants that have been bred. The F1 may have desired qualities of either or both parents.

- **F2**
  this is the product of two F1 plants that have been crossed. This is considered the second generation. This will not necessarily produce a great plant.

- **facilitated recurrent selection**
  a type of recurrent selection in which genetic male sterility is maintained in the population to maintain

hetero-zygosity and genetic diversity and to permit the recombination and shifting of gene frequencies.

- **facultative anaerobe**
  refers to an organism that normally grows aerobically but can grow anaerobically.

- **facultative organism**
  organism that can carry out both options of a mutually exclusive process (e.g., aerobic and anaerobic metabolism).

- **facultative parasite**
  an organism that normally lives as a saprophyte but under certain conditions can live as a parasite.

- **facultative saprophyte**
  a mainly parasitic organism with the ability to survive for a part of its life cycle as a saprophyte and be cultured on artificial media.

- **fairy ring**
  a circle of fungal growth.

- **falcate**
  of spores sickle-shaped

- **fallow**
  previously cultivated land kept free from crops or weeds during at least one growing season.

- **family**
  a group of plants, usually of several genera, and many species, which have the same basic floral structure and can thus be readily segregated and recognised from other families.

- **fasciation**
  this affects many herbaceous and woody plants. This is a genetic mutation or imbalance in growth caused by absorption of a herbicide. Remove all affected stems.

- **fasciculation**
  hyperplastic symptom characterised by a clustering of such plant organs as shoots into such structures as witches' brooms.

- **feed forward**
  the effect of process inputs (the levels of factors entering into a process) on the

process rate. In positive feed forward, the inputs amplify the rate, in negative feed forward, they diminish the rate.

- **feedback inhibition**
  inhibition by an end product of the biosynthetic pathway involved in its synthesis.

- **feeding value**
  1. characteristics that make feed valuable to animals as a source of nutrients.
  2. the combination of chemical, biochemical, physical, and organoleptic characteristics of forage that determine its potentials to produce animal meat, milk, wool, or work. Considered by some as synonymous with nutritive value.

- **fermentation**
  oxidation of certain organic substances in the absence of molecular oxygen.

- **fertigation**
  the application of fertiliser through an irrigation system.

- **fertilisation**
  the fusion of two gametes to form a new individual (zygote). Cross-fertilisation refers to male and female gametes from different flowers fusing.

- **fertilise**
  adding nutrients to the soil to promote growth and healthy plants - fertiliser, either artificial or natural substance that adds nutrients to the soil. Also the act of introducing pollen to produce seed.

- **fertiliser burn**
  the browning and in extreme cases, killing of plants from exposure to excessive nitrogen.

- **fescue foot**
  red and swollen skin at junction of hoof of animals grazing tall fescues, followed in advanced stages by gangrenous sloughing off of hoofs, tail tips, and ear tips, along with loss of appetite and emaciation. Problem is most severe in cold, rainy, or overcast weather.

- **fescue toxicity**
  1. exposure of an animal to the endophyte-infected tall fescue plant, eliciting a toxic response in the animal.

2. the state of being toxic to the animal.

- **fibre**
  1. a long, thick-walled cell (e.g., in cotton) serving to strengthen an organ.
  2. a unit of matter characterised by a length at least 100 times its diameter or width.

- **fibre fineness**
  a relative measure of size, diameter, linear density, or mass per unit length of fibre, expressed in a variety of units.

- **fibre length distribution**
  a graphic or tabular presentation of the proportion or percentage of fibres having different lengths.

- **fibre, textile**
  a generic term for the various types of matter that form the basic elements of textile fabrics and other textile structures.

- **fibrograph test beard**
  the portion of a test specimen (e.g., of cotton) that has been combed into a beardlike shape and that protrudes from the outside of a pair of fibrograph combs.

- **fibrous**
  a fine and sometimes dense root system.

- **fibrous root**
  a root system which contains many thin roots rather than a single tap root.

- **field burning**
  burning plant residue after harvest to aid in insect, disease, and weed control, reduce cultivation problems, and stimulate subsequent re growth and tillering of perennial crops.

- **field capacity**
  content of water, on a mass or volume basis, remaining in a soil after being saturated with water and after free drainage is negligible.

- **field grown**
  grown in the field, as opposed to root cuttings which are grown in pots in greenhouses.

- **filament**
  the part of the stamen that holds the anther in position for pollen dispersal.

- **filamentous**
  in the form of very long rods, many times longer than wide (for bacteria), in the form of long branching strands (for fungi).

- **filiform**
  thread-shaped.

Filiform (threadlike)

- **fimbria**
  short filamentous structure on a bacterial cell, although flagella-like in structure, generally present in many copies and not involved in motility. Plays a role in adherence to surfaces and in the formation of pellicles. (plural, fimbriae)

- **fine fescues**
  a fine-leaved turf grass that grows well in shade, low soil moisture, low fertility and low pH. It requires well-drained, slightly dry soils. Red, hard and chewing fescues are included in this group.

- **fines**
  material that passes through a screen whose openings are smaller than the specified minimum size of the product being processed.

- **fission**
  type of cell division in which overall cell growth is followed by formation of a cross wall which typically divides the fully grown cell into two similar or identical cells.

- **fistula**
  a surgical opening, duct, or passage from a cavity or hollow organ of the body.

- **fitness**
  the ability of an organism to survive and reproduce, the ability of an organism to

pass its genes to the next generation.

- **flagellate**
  protozoan that moves by means of one to several flagella.

- **flagellum**
  a whip-like appendage responsible for motility in the majority of motile bacteria and other protists, fungi, algae, etc. (pl. flagella).

- **flail mower**
  a mower that cuts turf by impact of blades rotating in a vertical cutting plane relative to the turf surface.

- **flask**
  a glass container used in the germination of orchid seeds and new seedlings.

- **flat**
  a shallow wooden box or plastic tray used to start cuttings or seedlings. Annuals may be purchased in a flat.

- **flexure**
  a turn or fold.

- **flint**
  maize variety with very hard kernels.

- **flora**
  the vegetation or plant life of a given region.

- **flore pleno**
  a botanical term describing a flower with extra petals.

- **floret**
  a small flower which is part of a much larger compound flower head, e.g. cineraria.

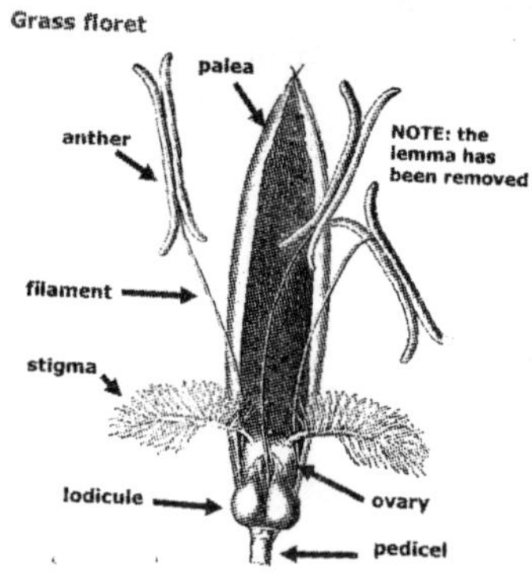

- **floricanes**
  on raspberries and blackberries, two-year-old canes which bear fruit and then die.

- **floriferous**
  free-flowering, easily brought into flower.

- **flower**
  the part of the plant with reproductive parts usually with petals. A double flower with one or more rows of petals.

- **flower spike**
  a flower head made up of a central stem with the flowers growing directly on it.

- **flowering stage**
  the physiological stage of a grass plant in which anthesis (blooming) occurs, or flowers are visible in non grass plants.

- **fluorescent**
  able to emit light of a certain wavelength when activated by light of a shorter wavelength.

- **fluorescent antibody**
  antiserum conjugated with a fluorescent dye, such as fluorescein or rhodamine.

- **flush**
  growth that is produced during a short period.

- **fluxes**
  rate of emission, sorption, or deposition of a material from one pool to another. For example, the exchange of methane between the land and the atmosphere is a flux, while the production of methane within the soil is not.

- **fodder**
  coarse grasses such as corn and sorghum harvested with the seed and leaves green or alive, then cured and fed in their entirety as forage.

- **foetid**
  With a disagreeable odour.

- **foiliar fertiliser**
  a liquid, water soluble, fertiliser applied to a plant's

foliage in a fine spray so that the plant can absorb the nutrients through its leaves.

- **foliaceous**
  leaf-like, used particularly in reference to sepals or bracts which simulate small or large leaves in texture, size, or colour.

- **foliar burn**
  injury to shoot tissue caused by dehydration due to contact with high concentrations of chemicals (e.g., certain fertilisers and pesticides).

- **forb**
  any herbaceous broadleaf plant that is not a grass and is not grass like. See also **legume**.

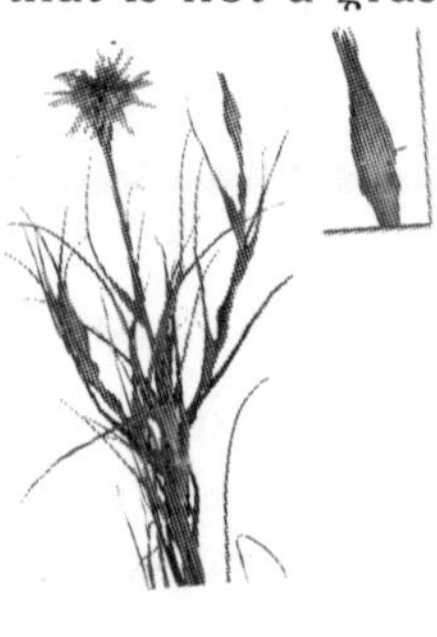

- **force**
  controlling the environment, e.g., heat, to produce early flowers or fruit.

- **forcing**
  the process of making a plant grow or flower before its natural season. Usually done indoors (i.e. paper whites).

- **forestland**
  land on which the vegetation is dominated by forest or, if trees are lacking, the land bears evidence of former forest and has not been converted to other vegetation.

- **form**
  an infraspecifc taxon in botanical nomenclature, below the rank of variety.

- **form genus**
  a non-phylogenetic category, equivalent to genus, distinguished on the basis of one or more morphological features. In the Deuteromycotina, form genera are used to classify anamorphs, such form genera are based primarily on

the characteristics (including mode of development of the conidia, conidiophores, and conidiomata.

- **form species**
  a non-phylogenetic category, equivalent to species, distinguished on the basis of one or more morphological features of an anamorph, treated as if it were an independent entity, especially for indexing or identification purposes, of importance chiefly in the Ascomycetes and Uredinales.

- **forma specialis**
  literally 'special form'. An infraspecific taxonomic rank in which the taxa are distinguished on a physiological basis, particularly on the basis of adaptation to (or pathogenicity) for one or more specific hosts. In mycology, forma specialis is a taxonomic rank lower than form, subvariety, variety and subspecies, and higher than physiological race.

- **forward creep**
  a method of creep grazing in which dams and offspring rotate through a series of paddocks with offspring as first grazers and dams as last grazers. A specific form of first-last grazing.

- **foundation planting**
  any plant that is used around a building for the sole purpose of making it look more aesthetic. In earlier days it was to cover the foundation.

- **foundation seed**
  foundation seed is the progeny of breeder, select, or foundation seed handled to maintain specific genetic purity and identity. Production must be acceptable to the certifying agency.

- **fpasture, rotation**
  pastureland used for a few seasons and then ploughed for other crops.

- **fractionation**
  separation into components, as by distillation, crystallisation, or physical separation.

- **free energy**
  intrinsic energy contained in a given substance that is available to do work, particularly with respect to chemical transformations, designated ÄG.

- **frequency of clip**
distance of forward travel between successive cuts of mower blades.

- **friable**
easily worked good, crumbly soil.

- **fringed**
furnished with hair-like appendages on the margins.

- **frond**
a leaf of a fern or palm. The limbs of a palm tree.

- **frontal grazing**
a grazing method that allocates forage within a land area by means of a sliding fence that livestock can advance to gain access to ungrazed forage.

- **frost**
the freezing and condensation of moisture in the air. Frost dates are important to know for your zone or area.

- **frost hardy**
plants that are able to survive winter frosts without damage to their leaves (i.e. evergreens) or damage to dormant stems, buds or roots (i.e. deciduous plants). Very much relative to geographic areas.

- **frost pocket**
a small area, often with lower ground than its surroundings, that is subject to lower temperatures making it particularly likely to get frosts.

- **frost tender**
these plants will be damaged or killed by even the lightest of winter frosts. Most evident would be tropical plants stretched to a cooler zone.

- **fructification**
1. synonymous with fruiting body.
2. the formation or development of a fruiting body.

- **fruit fly**
a small insect pest that will lay its eggs beneath the surface of developing fruits. The larvae will then grow quickly and exit through holes in the fruit or vegetable causing rot.

- **fruiting body**
macroscopic reproductive structure produced by some fungi, such as mushrooms, and some bacteria, including myxobacteria. Fruiting bodies are distinctive in size, shape, and colouration for each species.

- **fruiting wood**
on grapevine, the one-year-old canes that will produce the current year's fruit.

- **frustule**
siliceous wall and protoplast of a diatom.

- **fugacious**
withering quickly, falling off soon after anthesis (in reference to a flower).

- **full (rumen content)**
amount of ingested feed or water present in the rumen.

- **full shade**
this shade is sometimes called deep shade and is created by mature trees.

- **full sun**
six hours or more in the direct sun during the growing season of the year.

- **fulvic acid**
yellow organic material that remains in solution after removal of humic acid by acidification.

- **fumigant**
a gas or volatile substance that is used to disinfest certain areas of various pests.

- **fungi**
saprophytic and parasitic organisms that lack chlorophyll and include moulds, rusts, mildews, smuts, mushrooms and yeast (singular, fungus.)

- **fungi imperfecti**
see **deuteromycotina**.

- **fungicide**
a chemical or physical agent that kills or inhibits the growth of fungi. Note some substances termed 'fungicides' are fungistatic in their action.

- **fungistasis**
suppression of germination of fungal spores or other resting structures in natural soils as a result of competition for available nutrients, presence of inhibitory compounds, or both.

- **fungistat**
a substance that prevents fungal growth without killing the fungus.

- **fungus**
too vague to be an acceptable term for the fungus *Acremonium coenophialum* living symbiotically in tall fescue. Similarly, 'fungal absence', 'fungal incidence', 'fungal presence', 'fungus-free', and 'fungus-infected' are too vague without the context clearly specified.

- **furrow**
a depression in the planting garden either dug by a spade or a plow. It is created to be planted in or to be drainage.

- **fusarium**
this fungal disease is soil borne and causes wilting and death mostly in herbaceous plants.

- **fusarium solani**
macroconidia dimensions 27-52 x 4.4-6.8, Microcondia dimensions 8-16 x 2-4 microns. Found in plants and soils. Can produce trichothecene toxins which may be associated with disease in humans and animals.

- **fusiform**
spindle-shaped, tapering at each end, cigar-shaped.

- **gall**
an abnormal plant structure formed in response to parasitic attack by certain micro-organisms (bacteria, fungi, viruses or insects). Galls may develop either by

localised cell proliferation or increase in cell size.

- **gametangium**
  fungal structure that contains one or more gametes.

- **gamete**
  a haploid cell or nucleus involved in sexual reproduction, during which two gametes fuse to form a zygote.

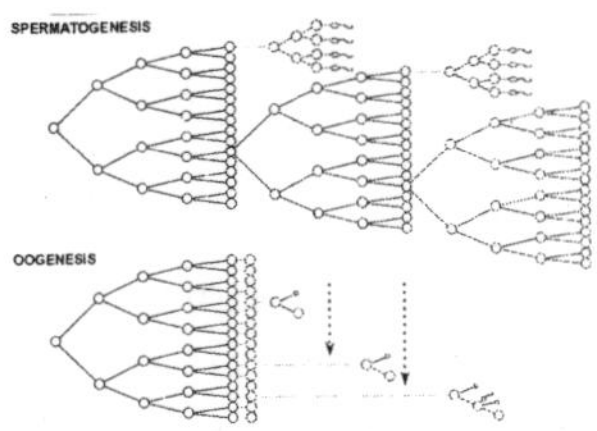

- **gametocide**
  an agent used to selectively kill either male or female gametes.

- **gametogenesis**
  the development of gametes.

- **gametophyte**
  the phase of a life cycle which has half the normal number of chromosomes.

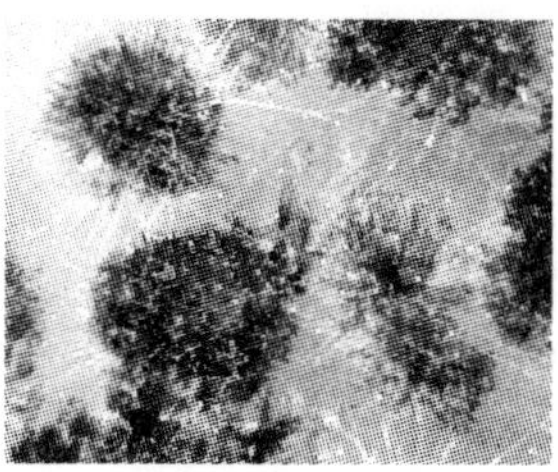

- **garden designer**
  a person who professionally will create plans for a home or public space. Many are self-taught, and not generally licensed by the state. Not to be confused with garden architects who are specifically trained and licensed. Garden designers sometimes are referred to as landscape designers.

- **gas chromatography**
  chromatographic technique in which the stationary phase is a solid or an immobile liquid and the mobile phase is gaseous. The gaseous samples are separated based on their differential adsorption to the stationary phase.

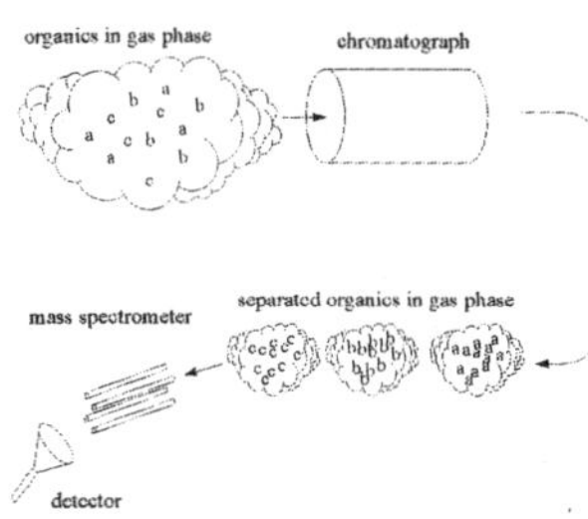

- **gasteromycete**
  a member of the Gasteromycetes.

- **gasteromycetes**
  the group of Basidiomycetes with spores borne in cavities within the fruit body.

- **gel**
  inert polymer, usually made of agarose or polyacrylamide, that separates macromolecules such as nucleic acids or proteins during electrophoresis.

- **gel electrophoresis**
  a type of electrophoresis in which the molecules in a sample moves through a gel composed of agarose or polyacrylamide.

- **gel filtration**
  a type of fractionation procedure in which molecules

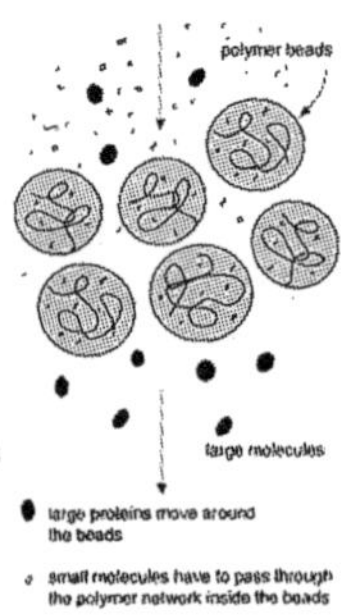

are separated from each other according to differences in size and shape, the action is similar to that of molecular sieves. Dextran gels (three-dimensional networks of polysaccharide chains) are usually used in this method.

- **gemini virus**
  from Latin Gemini, 'twins', for the typical double particles. Member of the only group of plant viruses with double particles, each particle containing circular or linear single-stranded DNA, this group is divided into two subgroups I and II (transmitted in a circulative manner by leafhoppers) and subgroup III (transmitted in a circulative manner by whiteflies) or, in the case of some viruses in this group, mechanically transmitted.

- **gene**
  unit of heredity, a segment of DNA specifying a particular protein or polypeptide chain, a TRNA or an MRNA.

- **gene cloning**
  isolation of a desired gene from one organism and its incorporation into a suitable vector for the production of large amounts of the gene.

- **gene expression**
  the transcription of MRNA from the DNA sequence of a

gene and the subsequent translation of that MRNA to give the protein gene product. Less strictly it can mean the transcription step alone.

- **gene probe**
  a strand of nucleic acid which can be labelled and hybridised to a complementary molecule from a mixture of other nucleic acids.

- **generation time**
  time needed for a population to double in number or biomass.

- **generic**
  of or pertaining to a genus.

- **genetic code**
  information for the synthesis of proteins contained in the nucleotide sequence of a DNA molecule (or in certain viruses, of an RNA molecule).

UUU UUC phenylalanine UUA UUG leucine	UCU UCC UCA UCG serine	UAU UAC tyrosine UAA UAG stop	UGU UGC cysteine UGA stop UGG tryptophan
CUU CUC CUA CUG leucine	CCU CCC CCA CCG proline	CAU CAC histidine CAA CAG glutamine	CGU CGC CGA CGG arginine
AUU AUC AUA isoleucine AUG methionine	ACU ACC ACA ACG threonine	AAU AAC asparagine AAA AAG lysine	AGU AGC serine AGA AGG arginine
GUU GUC GUA GUG valine	GCU GCC GCA GCG alanine	GAU GAC aspartic acid GAA GAG glutamic acid	GGU GGC GGA GGG glycine

- **genetic correlation**
  correlation between the genotypic values of two characters, or of two individuals with respect to the genetic character or different characters.

- **genetic engineering**
  in vitro techniques for the isolation, manipulation, recombination, and expression of DNA.

- **genetic shift**
  a change in the germ plasm balance of a cross-pollinated variety, usually caused by environmental selection pressures.

- **genome**
  the genetic information for an organism, consisting (in the case of viruses of one or more species) of either RNA or DNA, but not both.

- **genotype**
  the genetic constitution of an organism.

- **genus**
  used when naming plants. Genus is the plant equivalent of our surnames. When followed by the name of the 'species' you have it's botanical name. Almost always in Latin.

- **geographic range**
  geographic limits of the ecologic range, geographic extent of actual occurrences of a species.

- **geotrichum**
  conidia dimensions 6-12 x 3-6 microns. Aw 0.90 A common contaminant of grains, fruits, dairy products, paper,

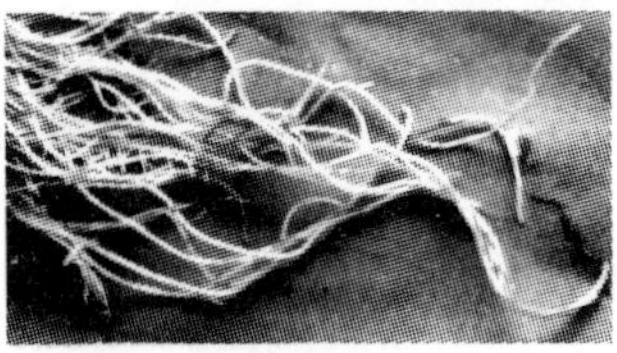

textiles, soil and water , and often present as part of the normal human flora. The species *Geotrichum candidum* can cause a secondary infection (geotrichosis) in association with tuberculosis. This rare disease can cause lesions of the skin , bronchi, mouth, lung and intestine.

- **geotropism**
  not a commonly used term but it is the response to gravity. Plant parts that grow downward, such as the roots, would be positive geotropism. A negative geotropism would be the stems growing upward.

- **germ tube**
  a short, hypa-like structure that develops from certain types of spores upon germination.

- **germinate**
  the sprouting of a seed.

- **germination**
  1. resumption of active growth by the seed embryo, culminating in the development of a young plant.

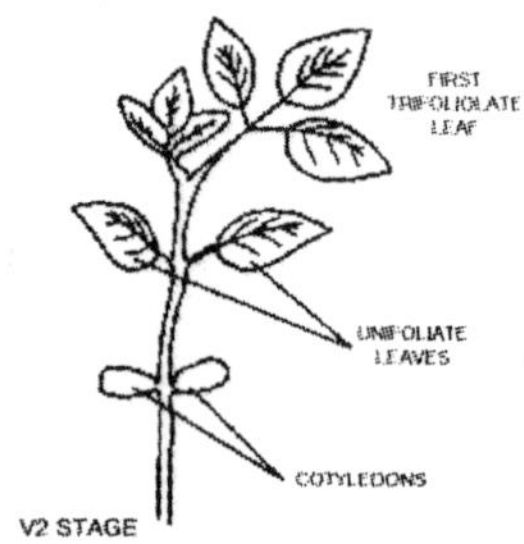

  2. in seed laboratory practice: emergence and development from the seed embryo of those essential structures, which, for the kind of seed in question, are indicative of the ability to produce a normal plant under favourable conditions.

- **germination by repetition**
  on the germination of a spore, the formation of a secondary spore rather than a germ tube.

- **gibberellin**
  a hormone used in plant production. Often used in Camellia blooms and in increasing the size of fruits.

- **gibb's free energy**
  see **free energy**.

- **gill**
  a hymenium-covered, plate-like appendage that hangs from the under surface of the basidiocarp of some Hymenomycetes.

- **ginned lint**
  cotton fibres after they have been removed from the seed.

- **girdling**
  the choking of a branch by a wire, rope, or other inflexible material which usually occurs most often in woody stemmed plants that have been tied down too tightly without regard for growth.

- **glabrous**
  plant surface which is smooth and hairless.

- **glade**
  an open space in a woodland area.

- **glandular**
  with glands, secreting organs, often tiny, which usually make the plant sticky.

- **glaucous**
  plant surface which is covered with a bluish-grey bloom.

- **glen**
  a very romantic term meaning a narrow valley.

- **gliocladium**
  a fungus which is structurally similar to penicillium sp. It is reported to be allergenic.

- **globose**
  spherical.

- **glochid**
  tiny, still hairs with barbs found in cacti.

• **glutenin**
the dilute alkali-soluble fraction of proteins in wheat.

• **glycolysis**
reactions of the Embden-Meyerhof (glycolytic) pathway in which glucose is oxidised to pyruvate.

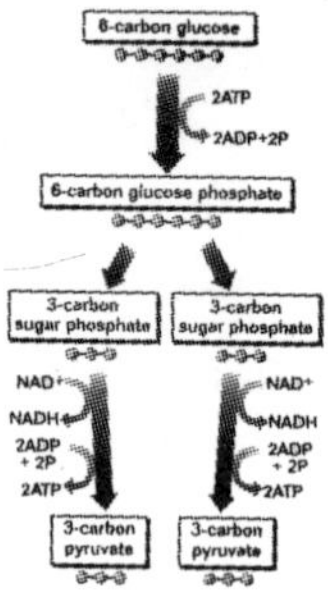

• **glycosidase**
enzyme that hydrolyses a glucosidic linkage between two sugar molecules.

• **grade**
not your A, B, C's in class, but the degree or direction of a slope, generally. Real important with house construction and ground placement.

• **grade, cotton**
a system of scales used in classifying cotton with regard to colour, trash content, and preparation.

• **graft**
a method of joining plants together, for example fruit trees are often grafted on to a more vigorous rootstock.

• **grafting**
this is a method of propagation. The process of joining a desirable stem or bud of

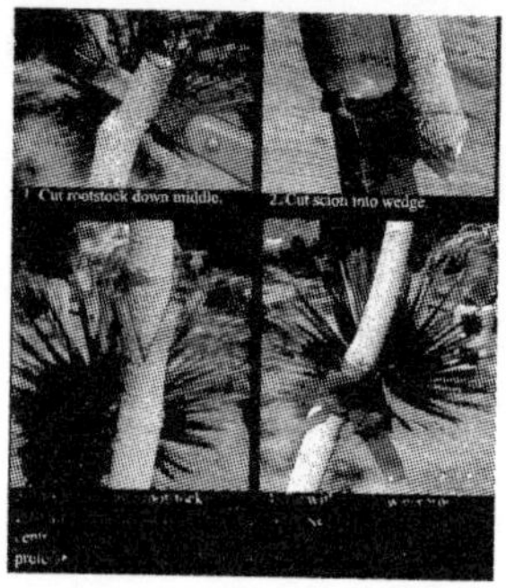

one plant (known as the scion) on to the less desirable, but hardier, stem of another (known as the stock). This will give a stronger root system than the scion would have normally had. Commonly done in roses, fruit trees and in some ornamentals.

• **grain grade**
market standard established to describe the amount of contamination, grain damage, immaturity, test weight, and marketable traits.

- **grain, percent**
threshed grain weight x 100/ (threshed grain weight + stalk weight).

- **grain-to-stalk ratio**
threshed grain weight/stalk weight.

- **gram reaction**
the result of the gram stain.

- **gram stain**
differential stain that divides bacteria into two groups, Gram-positive and Gram-negative, based on the ability to retain crystal violet when decolourised with an organic solvent such as ethanol. The cell wall of Gram-positive bacteria consists chiefly of peptidoglycan and lacks the outer membrane of Gram-negative cells.

- **granular fertiliser**
a fertiliser that is dry and is a tiny pellet form. It is spread able and should be measured. A granular fertiliser can come in both a natural and synthetic form.

- **grass**
member of the plant family Poaceae.

- **grassland**
1. land on which the vegetation is dominated by grasses.
2. more generally, any plant community in which grasses and/or legumes make up the dominant vegetation.

- **gravitational water**
portion of total soil water potential due to differences in elevation.

Hydraulic Head Diagram

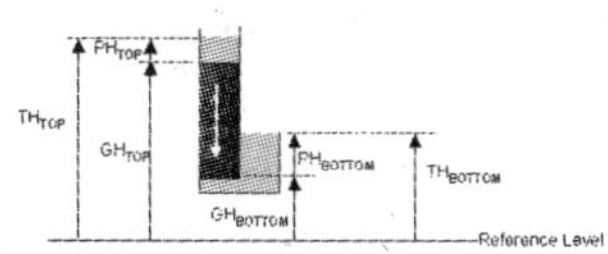

$TH_{TOP}=GH_{TOP}+PH_{TOP}$ $TH_{BOTTOM}=GH_{BOTTOM}+PH_{BOTTOM}$

$TH_{TOP}>TH_{BOTTOM}$ so water moves downward

- **grazable forestland**
forestland that produces, at least periodically, sufficient under-story vegetation that can be grazed. Forage is indigenous or, if introduced, is managed as though it were indigenous.

- **grazer**
  1. animal that grazes in situ grass as herbage.
  2. on experimental pastures, animals that may or may not remain on specified pasture treatment for the entire grazing period or season but are of a kind or physiological condition not necessarily represented on all pasture treatments for the entire grazing period or season.

- **grazing capacity**
  see **carrying capacity**.

- **grazing event**
  the length of time that an animal grazes without stopping.

- **grazing land management**
  the manipulation of the soil-plant-animal complex of the grazing land in pursuit of a desired result.

- **grazing management unit**
  the grazing land area used to support a group of grazing animals for a grazing season. It may be a single area, or it may have a number of subdivisions.

- **grazing period**
  the length of time that grazing livestock or wildlife occupy a specific land area.

- **grazing pressure**
  the relationship between the number of animal units or forage intake units and the weight of forage dry matter per unit area at a given time, an animal-to-forage relationship. The inverse of forage allowance.

- **grazing system**
  a defined, integrated combination of animal, plant, soil, and other environmental components and the grazing

method or methods by which the system is managed to achieve specific results or goals. Usage: Descriptive common names may be used, however, the first usage of a grazing system name in a publication should be followed by a description using a standard format. This format should include at least the following information: number, size, kind, slope, erosion status, and soil classification of land units, number, kind, sex, size, and age of livestock, duration of use and non use periods for each unit in the system, grazing methods, types of forage, geographic location and elevation, and type of climate and mean annual and seasonal temperatures and precipitation.

- **green manure**
  a crop (such as rye grass) that is grown and then incorporated into the soil to increase soil fertility or organic matter content. Usually turned over into the soil a few weeks before new planting begins.

- **greenhouse**
  a house that is green. A structure that can be built out of glass, plastic, or fibreglass. This building will be controlled in its temperature and humidity. Greenhouses are used for public display, cultivation, and in general protection of plants. Greenhouse comes in hobby sizes and as commercial usage.

- **greensand**
  a sediment composed of grains of glauconite mingled with clay or sand used as an organic fertiliser. It contains about twenty-two trace minerals including potash, silica, iron oxide, magnesia, lime, and phosphoric acid. Roses and tomatoes love it. The material will prolong fruiting and loosen heavy clay soils.

- **greenwood cutting**
  a cutting taken of young tips, after the first flush of growth in the spring has finished. They are somewhat hardier than softwood cuttings.

- **gregarious**

  growing together in clusters or colonies.

- **grex**

  a Latin word meaning 'group' or 'flock', the name used to describe a group of offspring of any given hybrid cross. When a grex name is registered, all additional identical crosses, plants produced from seeds of that cross or any asexual divisions of the cross all have the same grex name 'orchid hybrid'. The International Orchid Register is the century old international registration system for orchid hybrids. Its purpose is to ensure that grex nomenclature is uniform, accurate and stable, free from duplication and in accord with internationally agreed rules. The Orchid Review is the first place in which all new grex registrations are published for the first time, thus providing an important international service to the orchid world.

- **groat**

  the caryopsis or kernel of oat after the husk has been removed.

- **gross energy**

  the amount of heat that is released when a substance is completely oxidised in a bomb calorimeter containing oxygen at 25 to 30 atmospheres pressure.

- **ground cover**

  a plant used to provide a low-growing carpet between other plants.

- **groundwater**

  portion of the water below the surface of the ground at a pressure equal to or greater than atmospheric.

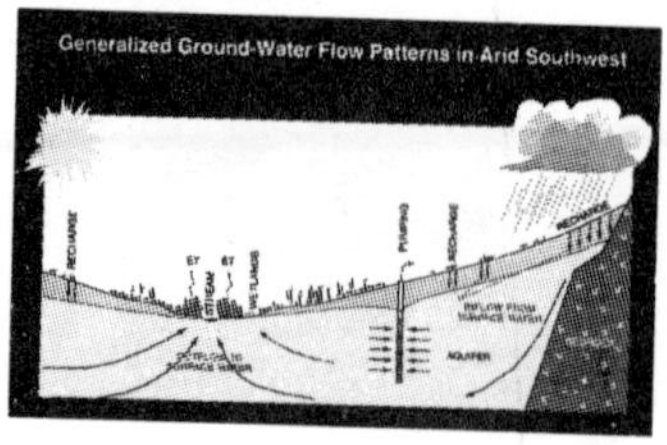

- **growing habit**
a direction or shape a plant takes as it grows.

- **growing point**
the area where the new growth occurs. When a plant is pinched and the new shoots then develop, this is the growing point.

- **growing season**
the period of time from the last frost date in spring to the first frost date in the fall. Vegetables especially will require a certain number of days to maturity.

- **growth**
in microbiology, an increase in both cell number and cellular constituents.

- **growth analysis**
mathematical analysis of crop or plant growth (following original concepts of Blackman and of West, Briggs, and Kidd) using Relative Growth Rate (RGR), Net Assimilation Rate (NAR), Leaf Area Growth Rate (LAR), and Crop Growth Rate (CGR). Generally involves the relationship of these variables to other crop or environmental variables as integrated over time (sampling interval).

- **growth factor**
organic compound necessary for growth because it is an essential cell component or precursor of such components and cannot be synthesised by the organism itself. Usually required in trace amounts.

- **growth rate**
a rate of change in growth (e.g., in number size or mass) of an organ, organism or community. Absolute growth rate is represented by dx/dt as change in amount per unit time. Specific (or relative) growth rate is dx/xdt, change per unit amount per unit time (with units of 1/time). Crop growth rate is a specific plant growth analysis term denoting the absolute growth rate of mass per unit land, A, thus dw/Adt, usually in units of grams per meter per day.

- **growth rate constant**
slope of $\log_{10}$ of the number of cells per unit volume plotted against time.

- **growth regulator**
a commercial chemical used by nurseries to change the shape of a plant. In general to dwarf a plant or to make the stems shorter. Kalanchoe are often dwarfed and then forced to bloom. The plant eventually will grow out of this, especially if cuttings are taken.

- **growth stage**
see **developmental stage**.

- **growth yield coefficient**
quantity of biomass carbon formed per unit of substrate carbon consumed.

- **grub**
short, fat, worm-like larva, especially of beetles.

- **guard cells**
specialised crescent-shaped cells that control the opening and closing of stomata.

- **gummosis**
a plant disease in which the lesions exude a sticky liquid.

- **guttation**
exudation of water from plants, particularly along the leaf margin.

- **gynostemium**
the technical term for the orchid's column.

Easy-care Cymbidiums flourish indoors in bright light or in a greenhouse where daytime temperatures average 65 degrees F to 85 degrees F. Cooler autumn temperatures trigger the flowering process.

- **gypsum**
a mineral of calcium sulphate. Gypsum adds calcium to the soil. It also will improve the structure of a clay soil. There will be no change in the pH value of the soil.

- **gypsy moth**
a caterpillar about 1 $^1/_2$ inches long that came from

Europe. These larvae do great damage by chewing and sometimes defoliating the entire tree.

- **habit**
the shape or form of a plant, growing vertical, laterally, or rounded. It is important to know the habit of a plant so one can expect certain growth patterns.

- **habitat**
the environment in which a plant is usually found growing, the factors being climate and soil.

- **half-hardy**
just as the word depicts, not completely hardy. Good examples are pansies, snapdragons, sweet alyssum. They will tolerate some light frost and cold nights.

- **halogen**
any of the five elements F, Cl, Br, I, and At that form part of group VII A of the periodic table.

- **halophile**
organism requiring or tolerating a saline environment.

- **halophyte**
those plants that will tolerate salt in the soil. Mostly coastal plants in the dunes, or marshes.

- **haploid**
in eukaryotes, an organism or cell containing one chromosome complement and the same number of chromosomes as the gametes.

Seksuell reproduksjon via diploid og haploid fase

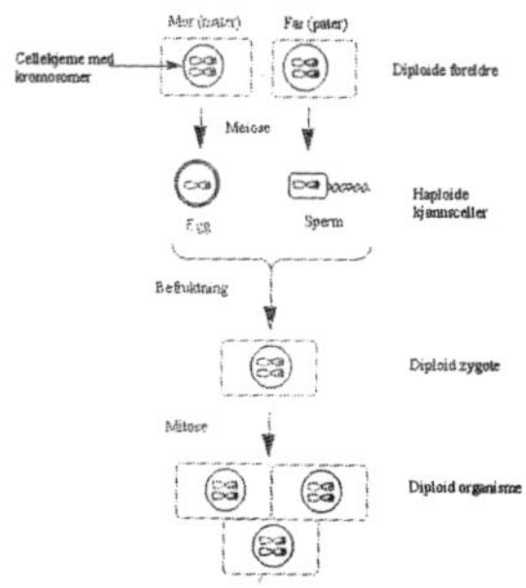

- **hard seed**
seed that is dormant due to a seed coat impervious to either water or oxygen.

- **harden off**
to acclimate a plant to harsher conditions.

- **hardening off**
  gradual acclimatisation to colder conditions. Usually used when taking seedlings out of the greenhouse or moving outside to a cold frame or protected area.

- **hardiness**
  capability of an organism to withstand environmental stress. Synonym: stress tolerance.

- **hardpan**
  compacted soil, sometimes new, created by construction. Water will run off and plant roots can not penetrate the layer. Can be broken up.

- **hardscape**
  includes any garden feature that is not a plant. Like birdbaths, deck, fences, trellises, benches, and patios.

- **hardwood cutting**
  a mature, woody piece of a woody plant that is removed to asexually propagate a new individual plant.

- **hardy**
  a plant which can withstand prolonged exposure to temperatures at or below 45° F.

- **hastate**
  spear shaped, with the basal lobe turned outward.

- **haustorium**
  a specialised branch of a hypha formed inside a host cell by certain plant-parasitic fungi especially obligate parasites in order to obtain nutrients. (pl. haustoria).

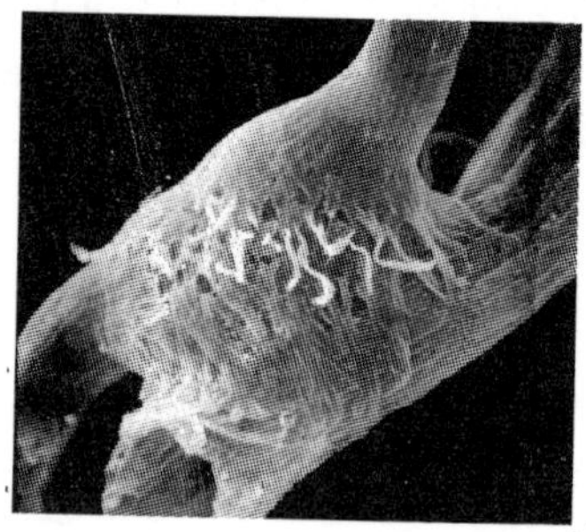

- **hay**
  stems of grass. Most popular feed for horses and cattle, however gardeners are always looking for 'spoiled hay' or 'mulch hay'. This is hay that has rotted or gotten mouldy. It is great for compost piles and using as a mulch. Beware of weed seeds and compost well.

- **head components**
components of the inflorescence of grains and grass crops, generally, the grain rather than vegetative structures, but may include individual vegetative structures such as rachis, peduncle, and pedicel.

- **heading**
the developmental stage of a grass plant from initial emergence of the inflorescence from the boot until the inflorescence is fully exerted.

- **heading back**
the process of cutting an older branch or stem to a twig or stub.

- **heart rot**
most commonly heard of in reference to trees. The centre merely has rotted out. Can also happen in root vegetables.

- **heave**
the partial lifting of a plant out of the soil as a result of alternating freezing and thawing of the soil.

- **heaving**
when there is a climate change from frost to warming of the soil, it often causes the soil to buckle upward. Sometimes called frost heaves.

- **heavy metals**
1. the heavy metals of concern to gardeners are lead, zinc, nickel, arsenic, copper and cadmium. These metals can be toxic to plants when they accumulate to high levels in the soil.
2. those metals which have densities more than 5.0 Mg m^{-3}. These include the metallic elements Cu, Fe, Mn, Mo, Co, Zn, Cd, Hg, Ni, and Pb. Al and Se have densities less than 5 but are also considered heavy metals.

- **hedge**
suitable trees, shrubs, or bushes planted relatively close together so that the branches will intertwine to provide a barrier fence for a windbreaker or privacy.

Hedges can be any height or width depending on the plant material used. Generally they are long lived species.

- **hedge row**
  a row of plant material (shrubs and trees) that are suitable for hedges.

- **heel**
  a strip of bark and wood remaining at the base of a side shoot cutting pulled off a main shoot. Some cuttings root more readily if a heel is attached.

- **heel cutting**
  a short, side branch taken as a cutting with a small piece of the main stem. (Often taken with old rose cuttings).

- **heeling in**
  1. covering the roots of dormant plants with soil or mulch for short periods.
  2. this is a temporary planting procedure until a plant can be put in its permanent place. The plant will benefit from the soil temperature.

- **heirloom plant**
  plants that have been around for 50 years or more. Not all people will consider the same plant an heirloom.

- **heirloom vegetables**
  cultivars that were popular a generation or more ago.

• **helicospore**
a cylindric, spiral or convolute spore, usually septate.

• **helminthosporium**
reported to be allergenic.

• **hemicellulose**
polysaccharides that accompany cellulose and lignin in the cell walls of green plants, differs from cellulose in that it is soluble in alkali and with acid hydrolysis gives rise to uronic acid, xylose, galactose, and other carbohydrates as well as glucose.

• **hemizonid**
in nematodes: Lens-like structure situated between the cuticle and hypodermal layer on the ventral side of the body just anterior to the excretory pore, generally believed to be associated with the nervous system.

• **hemizonion**
in nematodes: A companion structure to the hemizonid, it is smaller and located posterior to the hemizonid.

• **herb**
a plant grown for its medicinal or flavouring qualities, or its scented foliage.

• **herbaceous**
a plant with a non-woody stem. The upper parts will die back at the end of the growing season. It generally refers to perennials. Gertrude Jekyll invented the term.

• **herbaceous border**
probably more known in ferns with Gertrude Jekyll as she created great colourful perennial boarders in contrast to previous Victorian annuals in their showy and symmetrical beds. Of great interest and diversity.

• **herbage**
the biomass of herbaceous plants, other than separated grain, generally above ground but including edible roots and tubers.

• **herbarium**
a collection of dried specimens and a research centre. It is a special kind of museum.

- **herbicide**
any chemical that will kill a plant. There are both selective and non-selective herbicides. Selective only kill a specific plant and the non-selective kill a larger segment of plants.

- **herbivore**
a plant-eating animal.

- **heterocyst**
differentiated cyanobacterial cell that carries out dinitrogen fixation.

- **heteroecious**
requiring more than one host species to complete a life cycle (e.g., Uredinales)

- **heterogamy**
plasmogamy between morphologically different gametes.

- **heterogeneous**
characterising individuals with non-homozygous allelic genetic constitutions, or populations of plants with individual plants of different genetic constitution. Usage: Not applied to a single locus, where the term is heterozygous. n. heterogeneity.

- **heterokaryon**
a cell that contains genetically different nuclei or a thallus made up of such cells.

- **heterokaryosis**
the result of forming a heterokaryon of a fungus, the condition of a hypha or cell having two or more genetically distinct haploid nuclei.

Heterokaryon Formation in Fungi

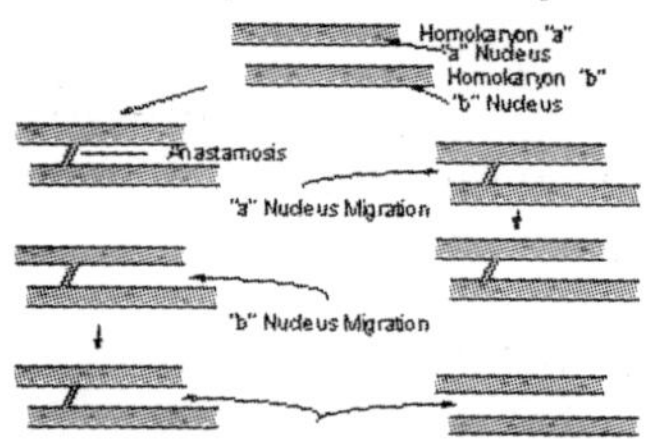

Two Heterokaryotic Hyphae

- **heterokaryotic**
the condition of being a heterokaryon.

- **heteroploid**
having a complement of chromosomes differing from that characteristic of the species.

- **heterothallic**
hyphae that are incompatible with each other, each requiring contact with another hypha of compatible mating type which, upon fusion, forms a dikaryon or a diploid.

- **heterothallism**
the phenomenon in which sexual reproduction requires the involvement of two different thalli.

- **heterotopy**
hyperplastic symptom in which an organ develops in a position other than its normal one.

- **heterotroph**
organism capable of deriving carbon and energy for growth and cell synthesis from organic compounds, generally also obtain energy and reducing power equivalents from organic compounds.

- **heterotrophic nitrification**
biochemical oxidation of ammonium to nitrite and nitrate by heterotrophic micro-organisms.

- **high shade**
always a difficult gauge but generally it is the shade beneath trees that have been 'limbed up' or pruned to get some light in.

- **high-intensity grazing**
usage: Not an acceptable term (it is a relative concept, best described in terms of management and method).

- **hill planting**
grouping plants in a cluster, not necessarily on an elevated mound.

- **hill up**
it sounds like an old fashion term but means to pull the soil around the stem of a plant for support.

- **histochemistry**
the chemistry of cells and tissues.

- **histopathology**
the study of pathology of cells and tissues, the microscopic changes characteristic of disease.

- **histoplasma**
a fungus which has filamentous growth at 25° C and yeast growth at 37° C. It is reported to be a human pathogen. It may be associated with birds.

- **holdfast**
a part of a plant that clings to a flat surface.

- **holobasidium**
a single-celled basidium. (pl. holobasidia)

- **holoblastic**
when both outer and inner walls of the conidiogenous cell contribute to the formation of the blastoconidium.

- **holocarpic reproduction**
in fungi, reproduction in which the entire fungal body is segmented into spores.

- **holomorph**
any fungus considered in its entirety, i.e., including all latent or expressed (anamorphic or teleomorphic forms).

- **holotype**
the single specimen designated or indicated as 'the type' by the original author at the time of publication of the original description.

- **homokaryon**
a hyphal cell, mycelium, organism, or spore in which all the nuclei are genetically identical.

- **homokaryotic**
refers to a homokaryon.

- **homonym**
a taxonomic designation rejected because the identical term has been used to designate another group of the same rank.

- **homothallic**
hyphae that are self-compatible in that sexual reproduction occurs in the same organism by meiosis and genetic recombination. Fusion of hypha results in a dikaryon or diploid.

- **honey dew**
the sweet and sticky syrup secreted by aphids and other sap-sucking insects. It then sometimes gets mouldy fungus on it. Should be washed off with a soapy water rinse.

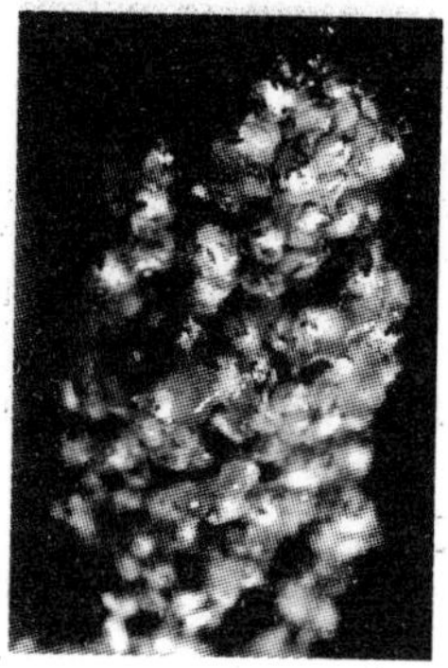

- **horizon**
see **soil horizon**.

- **horizontal resistance**
in a given cultivar: the existence of similar levels of resistance to each of the races of a given pathogen.

- **horticultural oil**
this includes both a dormant oil and a summer oil used

to smother eggs and developing insects on trees and ornamentals. The heavier oils are used in the late winter or very early spring, making sure the temperatures are over 40°F but before the plant leafs out. The lighter summer oil can be used anytime the temperature is below 85°F.

- **horticulture**
  the art and science of gardening. Commercial horticulture would include fruit, flower and small scale vegetable growing including the nursery industry. Agriculture covers broad acre farming of plants.

- **host**
  any plant material that will support a parasite. Oak trees will host mistletoe that will create damage and oaks can also host Spanish moss that does no harm.

- **hot cap**
  cones that are used a lot in cooler zones to protect newly planted seedlings from cold weather and birds. Usually made of plastic or a very heavy translucent waxed paper.

- **hot house**
  another term for a greenhouse sometimes found in earlier garden writings.

- **hot bed**
  a bed of soil enclosed by a structure with a top of glass, heated, often by manure, for forcing or raising seedlings.

- **hot-water insoluble nitrogen**
  fertiliser nitrogen not soluble in boiling-point hot water (100° C, 212° F), used to determine activity index of urea forms.

- **house plants**
  plants that are grown in containers inside the home.

- **humic acid**
  dark-coloured organic material extracted from soil by various reagents (e.g., dilute alkali) and that is precipitated by acid (pH 1 to 2).

- **humic substances**
  series of relatively high-molecular-weight, brown-

to-black substances formed by secondary synthesis reactions. The term is generic in a sense that it describes the coloured material or its fractions obtained on the basis of solubility characteristics, such as humic acid or fulvic acid.

- **humicuola**
grow on products with a high cellulose content. These fungi are also found in soil and on plant debris.

- **humidifier**
a piece of equipment used to raise the humidity of the air in a room.

- **humidity**
the amount of moisture in the air.

- **humification**
process whereby the carbon of organic residues is transformed and converted to humic substances through biochemical and chemical processes.

- **humus**
total of the organic compounds in soil exclusive of not decayed plant and animal tissues, their 'partial decomposition' products, and the soil biomass. The term is often used synonymously with soil organic matter.

- **hyaline**
transparent, translucent, or colourless.

- **hyaline mycelia**
sterile mycelia which is white or transparent. No fruiting structures are produced by the mycelia. Visual identification of these organisms is not possible. Often associated with allergic symptoms.

- **hybrid**
a plant which is the offspring of parents of different species. Hybrids are either intrageneric or intergeneric.

- **hybrid variety**
first-generation (F1) progenies from a cross, produced through controlling the pollination between two inbred lines, two single crosses, a single cross and an inbred line, an inbred line or a single cross and an open-pollinated or a synthetic variety, or two selected clones, seed lines, varieties, or species.

- **hybridisation**
natural formation or artificial construction of a duplex nucleic acid molecule by complementary base pairing between two nucleic acid strands derived from different sources.

- **hydathode**
a specialised leaf structure with one or more openings through which water is discharged from the interior of the leaf to its surface.

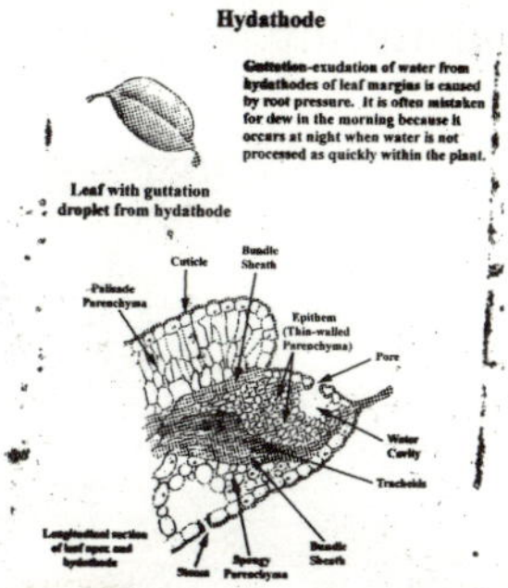

- **hydrocyanic acid**
HCN. A poison produced as a glucoside by several plant species, especially sorghums. Synonym: prussic acid.

- **hydrogen bond**
chemical bond between a hydrogen atom of one molecule and two unshared electrons of another molecule.

- **hydrogen-oxidising bacterium**
facultative lithotrophs that, in the absence of an oxidisable organic source, oxidise H_2 for energy and synthesise carbohydrates with carbon dioxide as their source of carbon.

- **hydroponics**
a method of growing a plant in water containing dilute

nutrients. Many vegetables are commercially grown this way. Epcot centre has a wonderful hydroponics centre.

- **hydroseeding**
planting seed in a water mixture by pumping through a nozzle that sprays the mixture onto a seed bed. The water mixture may also contain addends such as fertiliser and mulches.

- **hydrosis**
necrotic symptom of disease characterised by water-soaking of tissues (pl. hydroses).

- **hygrometer**
an instrument used to measure the Relative Humidity of the air.

- **hygroscopic water**
water adsorbed by a dry soil from an atmosphere of high relative humidity.

- **hymenium**
layer of hyphae which are fertile in producing asci (fungi in the phylum Ascomycota) or basidia (fungi in the phylum Basidiomycota) from the process of meiosis.

- **hymenomycetes**
the group of Basidiomycetes possessing an exposed hymenium.

- **hymenophore**
spore-bearing structure, the part of a basidioma bearing the hymenium.

- **hyperparasite**
a parasite of a parasite.

- **hyperplasia**
the enlargement of an organ or tissue owing to an increase in the number of cells.

- **hypersensitive**
the state of being abnormally sensitive. It often refers to an extreme reaction to a pathogen (e.g., the formation of local lesions by a virus or the necrotic response of a leaf to bacterial infection).

- **hypertrophy**
increase in cell size causing an increase in the size of an organ or tissue.

- **hyperturfa**
lately this has become very popular. A replica lightweight rock material created by equal amounts of dry cement, sand, perlite, peat moss, and water. This is made into troughs, planters, and rock all in the manner of looking old.

- **hypha**
filamentous part of a fungus, usually septate and consisting of several cells in linear succession. (pl. hyphae). See **mycelium**

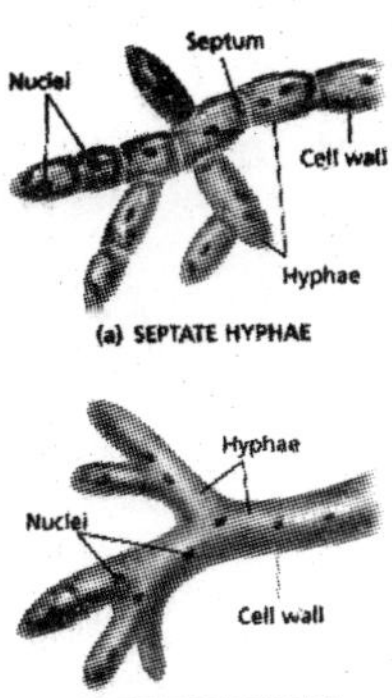

- **hyphal fusion**
see **anastomosis**.

- **hyphal peg**
1. a compound, hyphal, fasciculate projection extending beyond the general level of the hymenium, consisting of two or more parallel or interwoven hyphae, encrusted or gelatinised.
2. a projection from a hypha.

- **hyphomycetes**
a group of the deuteromycetes without differentiated pycnidia or acervuli.

- **hyphopodium**
a short branch of one or two cells of the epiphytic mycelium of a black mildew fungus (pl. hyphopodia).

- **hypochile**
lower or basal part of the lip in some orchids, as in Stanhopea.

- **hypogeal germination**
a type of germination in which the cotyledons remain below the ground while the epicotyl grows and emerges above the ground.

- **hypogeous**
growing below ground.

• **hyponasty**
more rapid growth of the lower side of an organ than of the upper side.

• **hypoplasia**
underdevelopment resulting from an abnormal paucity of cells.

• **hypovirulence**
a reduced level of virulence in a strain of pathogen resulting from genetic changes in the pathogen or to the effects of an infectious agent on the pathogen.

• **hysteresis**
a phenomenon evidenced by the fact that at a given relative humidity, seeds or grain may reach two different equilibrium moisture contents: one by increasing the relative humidity from a low level, and another by decreasing the relative humidity from a high level.

• **hysterothecium**
an ascocarp that opens by a slit.

• **idiomorphs**
nucleotide sequences that occupy the same locus in different strains but are not related in sequence or common descent

• **ikebana**
the formal flower arranging done by the Japanese. Special attention and thought to balance, harmony, and form.

• **illuminance**
see **light intensity**.

• **illuviation**
deposition of soil material removed from one horizon to another in the soil.

• **imbricate**
overlapping, like the shingles of a roof.

• **immobilisation**
conversion of an element from the inorganic to the

organic form in microbial or plant biomass.

- **immune**
  not subject to attack by a specified pest. Immunity is absolute.

- **immunity**
  1. the state of being immune
  2. in plants, the ability to remain free from disease because of inherent structural or functional properties.

- **immunoblot (western blot)**
  detection of proteins immobilised on a filter by complementary reaction with specific antibody.

- **immunofluorescence**
  technique to visualise specific antibodies and any attached homologous antigens by means of conjugating the antibodies to a fluorescent dye.

- **immunogen**
  substance which is capable of eliciting immune response. An immunogen usually has a fairly high molecular weight (usually greater than 10,000), thus, a variety of macromolecules such as proteins, lipoprotein, polysaccharides, and some nucleic acids can act as immunogens.

- **immunoglobulin**
  antibody.

- **imperfect fungus**
  see **deuteromycotina**.

- **imperfect stage**
  see **anamorph**.

- **imperfect state**
  the state of a fungus characterised by asexual spores (conidia) or the absence of spores.

- **in vitro**
  literally 'in glass', it describes whatever happens in a test tube or other receptacle, as opposed to in vivo. When a study or an experiment is done outside the living organism, in test tube, it is done in vitro.

- **in vitro dry matter disappearance**
  a gravimetric measurement of the amount of dry matter lost upon filtration following the incubation of forage in test tubes with rumen micro flora, usually expressed as a percentage: (Wt. dry matter sample - wt.

residue) / (weight dry matter sample).

- **in vivo**
  in the body, in a living organism, as opposed to in vitro. When a study or an experiment is done in the living organism, it is done in vivo.

- **in vivo nylon bag technique**
  system of determining dry matter disappearance of forage placed in fine-mesh nylon bags, either placed in the rumen or suspended in the rumen from a canula cover of a fistulated animal.

- **incidence**
  usage: Refers to the presence of a measured characteristic. In tall fescue studies, the term may (if properly defined earlier in a publication) be used to refer to, e.g., the number of infected plants or, if expressed as a percentage or in hundredths, to the rate of occurrence of a specified fungus in or on a known number of plants.

- **incipient**
  early in development (of a disease or condition).

- **incompatible**
  not cross-fertile.

- **incomplete metamorphosis**
  gradual growth of an arthropod that involves change in size, but not form.

- **incubation**
  the growth of a pathogen so that it can enter a host.

- **incubation period**
  the period of time between penetration of a host by a pathogen and the first appearance of symptoms on the host.

- **indehiscent**
  not splitting open at maturity, opposite of dehiscent.

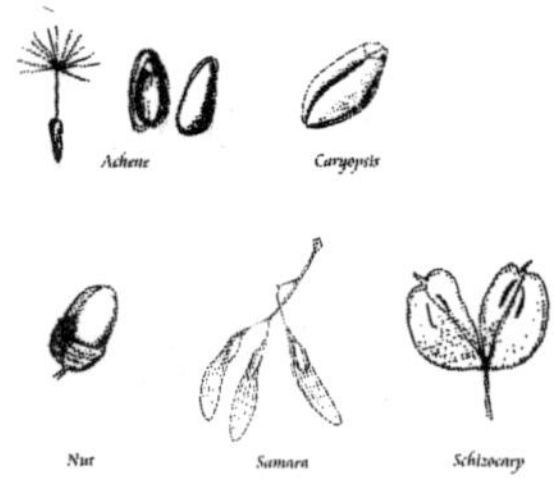

- **indeterminate**
  without definite margin or edge, terminal growth (of hyphae or conidiophores) unrestricted.

- **indeterminate inflorescence**
  flowers arising laterally and successively as the floral axis elongates.

- **indexing**
  a procedure to determine whether a given plant is infected by a virus. It involves the transfer of a bud, scion, sap etc. from one plant to one or more kinds of indicator plants sensitive to the virus.

- **indicator host**
  a plant species that gives characteristic symptoms to a specific virus. Used in virus diagnosis.

- **indigenous**
  plant species that are native to that region. These plants are believed to be growing in the wild at least for the span of human history.

- **inducible enzyme**
  enzyme synthesised (induced) in response to the presence of an external substance (the inducer).

- **infect**
  of a parasite: to begin or continue an interactive, usually pathogenic, relationship with the host.

- **infection court**
  a site in or on a host plant where infection can occur.

- **infection thread**
  the specialised hypha of a pathogenic fungus that invades tissue of the susceptible plant.

- **infectious**
  capable of producing propagules that disperse from one host to another and infect it.

- **infertile**
  soil that has no nutrients.

- **infest**
  of micro-organisms and viruses: to contaminate the surface of a plant without establishing an interactive relationship with it, to be present in high numbers in a plant's environment (soil, water, etc).

- **infrared**
  the portion of the electromagnetic spectrum with wavelengths from about 0.75 μm to 1 mm.

- **ingress**
  the act, by a plant pathogen, of gaining entrance into the tissues of a susceptible plant

- **inhibition**
  prevention of growth or function.

• **initiation**

1. inception of a new organ, as in organ formation at a meristem (e.g., leaf and floral part initiation at the shoot apical meristem), activation of an existing meristem (cambia in shoots and potential underground storage organs), or development of a new meristem (lateral root initiation, adventitious bud initiation).
2. a morphological change of the growing point from the vegetative condition to the floral condition.

• **innoculate**

to introduce a micro-organism into an environment suitable for its growth, to bring a parasite into contact with a host.

• **innoculum**

the population of micro-organisms introduced in an innoculation, the units of a parasite capable of initiating an infection.

• **inoperculate**

of an ascus or a sporangium, opening by a pore or split to discharge spores, as in asci of the Helotiales.

• **inorganic**

a chemical or fertiliser which is not obtained from a source which is or has been alive.

• **insecticidal soap**

a great alternative to using chemicals on plants. It may not be the total answer to getting rid of insects but certainly should be tried first. A homemade version is two tbsp. to a gallon of water. Several application may have to be used.

• **insecticide**

a synthetic or organic chemical used to kill or repel insects. Please use as directed on the label.

• **insectivorous plant**

another term for carnivorous plants. Plants that will trap and digest insects that

will supply them with nitrogen. Most of these plants

grow in swamps where it is difficult to obtain nitrogen.

- **insertion**
  genetic mutation in which one or more nucleotides are added to DNA.

- **insertion sequence (IS element)**
  simplest type of transposable element. Has only genes involved in transposition.

- **in situ**
  the act of sowing seeds or cuttings in the ground where they are to grow.

- **instar**
  the stage in the life of an arthropod between moults.

- **integration**
  process by which a DNA molecule becomes incorporated into another genome.

- **intercalary**
  formed or situated somewhere between apex and base of a given structure.

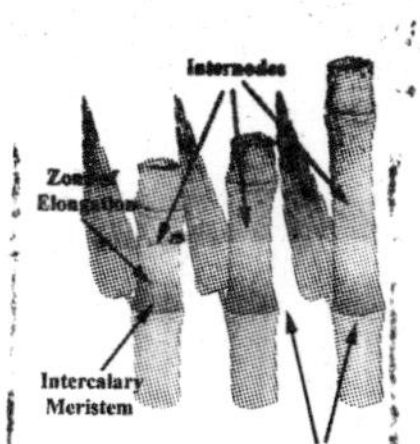

Intercalary Meristem
(only in monocot stem)

- **intercellular**
  between cells.

- **intercropping or interplanting**
  a smart way to garden. Mixing two or more plants, tall and short, for foliage difference, or combining plants that bloom at different times of the year. It keeps the interest in the garden.

- **intergeneric**
  term usually used when referring to Cross breeding of different species from different genera producing new hybrids. Genera are always genetically related members of the same taxonomic Tribe.

- **intermittent grazing**
  a method that imposes grazing for indefinite periods at irregular intervals.

- **internodal cutting**
  a cutting taken by cutting between two nodes.

- **internode**
  potion of a stem situated between the nodes or joints.

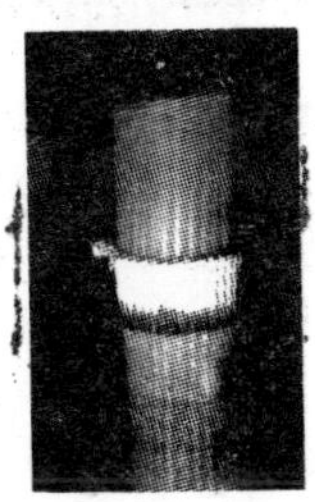

• **interplant**
growing two different intermixed crops in an area to maximise space usage.

• **interseeding**
seeding between sod plugs, sod strips, rows, or sprigs, or seeding into turf to improve the stand or alter its composition.

• **intersex**
an individual more or less intermediate in phenotype between male and female, displaying secondary male or female characters.

• **interstem**
an intermediate stem piece that is grafted between the scion and the stock.

Two-piece apple or pear tree (A) and interstem (B)

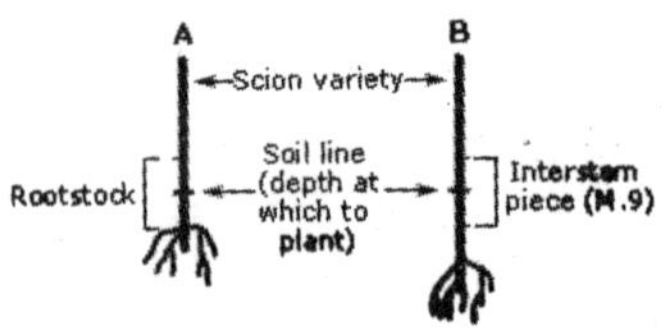

• **intracellular**
within or through the cells.

• **intrageneric**
term usually used when referring to cross breeding of different species of a single genera producing new hybrids.

• **introduced species**
a species not part of the original fauna or flora of the area in question, i.e., brought by human activity from another geographical region.

• **intumescence**
hyperplastic symptom characterised by blister-like swelling on the surfaces of plant organs.

• **invagination**
retraction, under force of pressure, of an outer surface toward the inside.

• **invasion**
in plant pathology: spread of a pathogen through tissues of a diseased plant.

• **invasive**
the ability of a plant to spread quickly and crowd out other plantings Great for a ground cover but dangerous for a well kept, under control garden bed.

• **involute**
of the margin of the pileus: rolled in, especially when young.

• **ions**
atoms, groups of atoms, or compounds, that are electri-

cally charged as a result of the loss of electrons (cations) or the gain of electrons (anions).

- **IPM**

Integrated Pest Management. A method by which gardeners can learn to manage and eradicate pests by choosing appropriate plants providing good growing conditions and minimising pests rather than annihilating them.

- **iron chelate**

this is often recommended when plants are showing signs of chlorosis. It is a form of iron, being readily absorbed by plants, especially in alkaline soils.

- **irrigation**

to supply water by artificial means, such as with sprinklers.

- **irrigation, automatic**

hydraulic-electric control of irrigation in response to a transducer that senses plant needs. The term is commonly used more loosely to refer to hydraulic or electrically actuated valves manually present on a time-based controller.

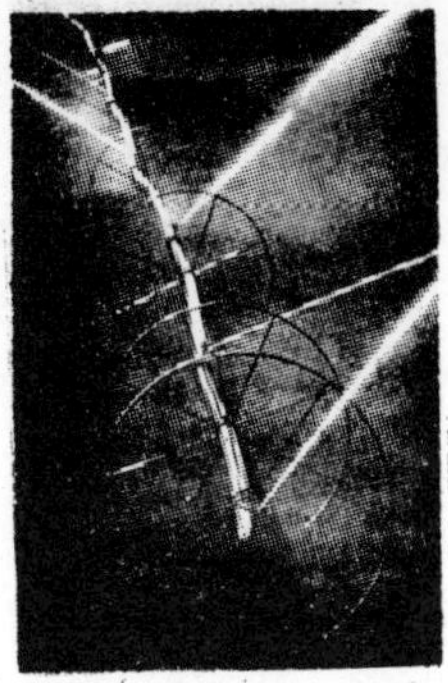

- **irrigation, semiautomatic**

an irrigation system in which valves respond directly to a manually operated remote-control switch.

- **isochromosome**

chromosome with arms equal in length and genetically identical, but with the loci in reverse order, produced when the centro mere splits transversely.

- **isoenzyme (isozyme)**

when two different enzymes catalyse the same reaction(s), they are isoenzymes of each other. Isoenzymes could differ from each other in their primary structure or electrophoretic mobility.

- **isogamy**

the condition in which gametes are morphologically

similar, as in the members of the Zygomycotina.

- **isogenic**
  two or more organisms or cells having identical genotypes.

- **isolate**
  in plant pathology: a culture or subpopulation of a microorganism separated from its parent population and maintained in some sort of controlled circumstance. Also, to effect such separation and control, for example, to isolate a pathogen from diseased plant tissue.

- **isolation**
  any procedure in which an organism present in a particular sample or environment, is obtained in pure culture.

- **isoline**
  isogenic line. One line in a series of genetically similar plant lines that carry different specific genes for resistance to a particular pathogen.

- **isomorphous substitution**
  substitution in a crystalline clay sheet of one atom by a similarly sized atom of lower valence.

- **isotope**
  different form of the same element containing the same number of protons and electrons, but differing in the number of neutrons.

- **isozyme**
  one of a number of enzymes that catalyse the same reaction(s) but differ from each other in primary structure and/or electrophoretic mobility. Also known as Isoenzyme.

- **isthmus**
  in nematodes: Relatively narrow portion of esophagus just anterior to the basal region.

- **Japanese beetle**
  its larvae in the soil is known as the white grub that will eat the roots of your grassy lawn. When the beetle appears it eats its way through plants especially roses and leaves them like skeleton forms.

Japanese Beetle

- **Japanese gardens**
  gardens that are designed with a Japanese cultural influence. Using particular plants for the design. Often with a Zen influence. Bamboo, pine, mondo grasses, koi are often used.

- **John Innes compost**
  a loam based compost.

- **juvenile**
  the life stage of a nematode between the embryo and the adult, an immature nematode.

- **juvenillody**
  condition in which tissues and organs remain immature.

- **karyogamy**
  fusion in a cell of haploid (N) nuclei to form a diploid (2N).

- **karyotype**
  the chromosomal constitution of a eukaryotic cell in terms of the number, size and morphology of the chromosomes at metaphase.

- **keel**
  a boat-shaped structure formed by the two lower petals of many members of the Leguminosae.

- **keiki**
  Hawaiian term used by orchidists to signify an offshoot or offset from a plant.

- **kentucky bluegrass**
  a cool-season turf grass that spreads by rhizomes. It is the most popular species for high-quality lawns in Ohio. It is very winter hardy.

- **kernel**
  the edible part of nut.

- **ketosis**
  a pathological accumulation of ketone bodies in the body.

- **klendusity**
  a special kind of disease escape in which a susceptible plant avoids disease because of an intrinsic property of the plant itself that greatly reduces the chances of its being innoculated, even though there may be an abundance of innoculum in the area.

- **knocking out**
the temporary removal of a plant from its pot in order to check the condition of the root ball.

- **knot**
a localised abnormal swelling, a gall.

- **knot garden**
a very carefully planned garden of small dwarf shrubs or even herbs. Planned in a pattern and kept in order by constant pruning and trimming.

- **koch's postulates**
set of laws formulated by Robert Koch to prove that an organism is the causal agent of disease.

- **k-strategy**
ecological strategy where organisms depend on physiological adaptations to environmental resources. K strategists are usually stable and permanent members of the community.

- **labellum**
lip, particularly that of an orchid.

- **labial**
in nematodes: Pertaining to the lips.

- **lacewing**
think beneficial About one inch in length will eat mites, aphids, and thrip.

- **lag phase**
period after innoculation of fresh growth medium during which population numbers do not increase.

- **lamella**
1. a thin layer, plate like arrangement or membrane.
2. layers of protoplasmic membranes within the chloroplast that contain the photosynthetic pigments. plural, lamellae.

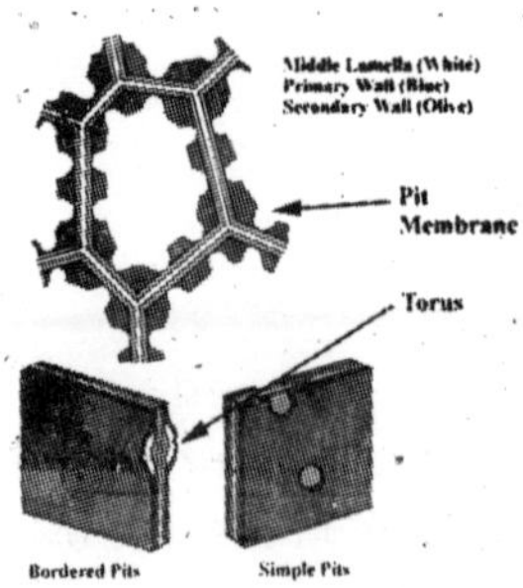

- **landscape**

a collective term for all the natural features (such as fields, hills, forests, water, etc.) that distinguish one part of the earth's surface from another part. Usually used in reference to that land or territory which the eye can comprehend in a single view, including all its natural characteristics.

- **landscape architect**

a very good professional to consider when creating a garden. They are licensed by the state and know the mechanics of major construction, grading, drainage, and will advise solutions to problems.

- **landscape fabric**

probably one of your best solutions against weeds for beds. This can be a variety of materials: newspaper, plastic, porous fibres, burlap. It provides a barrier to weeds germination. This is then dressed with mulch of any kind.

- **larva**

a stage of insect metamorphosis between the egg and pupal stages. The feeding, growing, non-reproductive stage of insect development.

- **latent bud**

a dormant bud that is capable of growth and development.

- **latent infection**

infection unaccompanied by visible symptoms.

- **latent period**

1. the period between infection and the appearance of new innoculum.
2. elapsed time between phage infection or induction and lysis of bacterial cells.
3. period after acquisition of virus by a vector before it becomes infective.

- **latent virus**

a virus that infects a plant without causing macroscopic symptoms.

- **lateral bud**

smaller buds on the sides of stems, responsible for growth of leaves and side branches.

- **lateral shoot**

shoots originating from vegetative buds in the axils of leaves or from the nodes of stems, rhizomes, or stolons.

- **latex**
  milky sap which exudes from cut surfaces of a few house plants, such as Ficus elastica decora and Euphorbia.

- **lath**
  a structure used to create shade. Made of any material, evenly spaced to let in some light.

- **lath house**
  a structure consisting of a frame supporting strips of wood which are spaced to provide about 50% shade.

- **latin binomial**
  see **binomial nomenclature**.

- **layering**
  a way to propagate plant material. A branch is anchored to the ground (still attached to the parent plant). Roots appear creating another plant.

- **layering, soil**
  stratification within a soil profile, which may affect conductivity and retention of water, soil aeration, and rooting. Can be due to construction design, topdre-ssing with different textured amendments, inadequate on-site mixing of soil amendments, or blowing and washing of sand or soil.

• **leach**
to dissolve in water and wash away.

• **leaching**
a similar concept to making tea which leaches out the flavour of the tea leaves. This concept regards how water will rinse bad substances (like salt) or good ones (like nutrients) down deep into the soil or as runoff.

• **leader**
the main growing shoot of a sapling, it eventually will become the trunk of the tree.

• **leaf bud cutting**
a cutting taken from a small piece of stem with one or two leaves or buds.

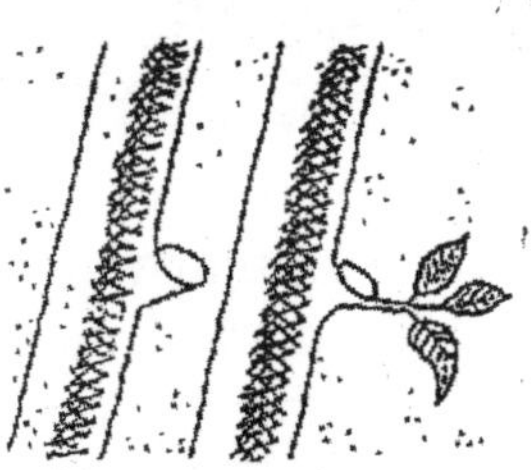

A Dutch cut encourages growth from below the cut

• **leaf cutting**
a method of propagation. A leaf is removed and is placed in vermiculite or perlite for future baby plants.

• **leaf mould**
partially decayed leaves used in some potting mixtures. It must be sieved and sterilised before use.

• **leaf roller**
different moths that will roll leaves when larvae. Cannas often get a lot of leaf rollers.

• **leaf scorch**
injury to leaves due to lack of sufficient water, excessive transpiration or injury to the water-conducting system of the plant.

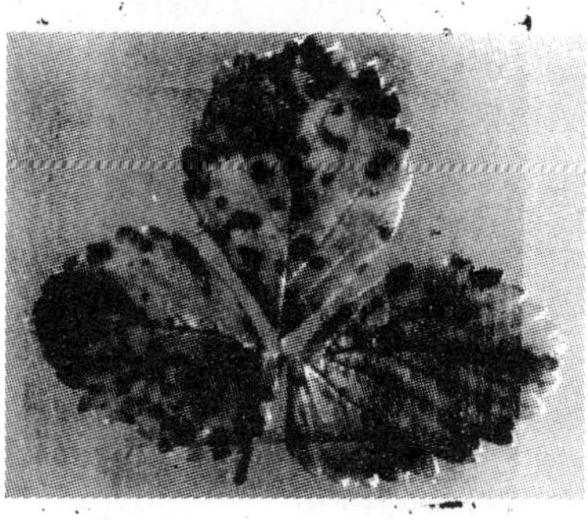

• **leaf spot**
a self-limiting lesion on a leaf.

- **leaf, early**
  a developmental stage, requires a reference point to be useful.

- **leaflet**
  a leaf-like section of a compound leaf. Small in stature.

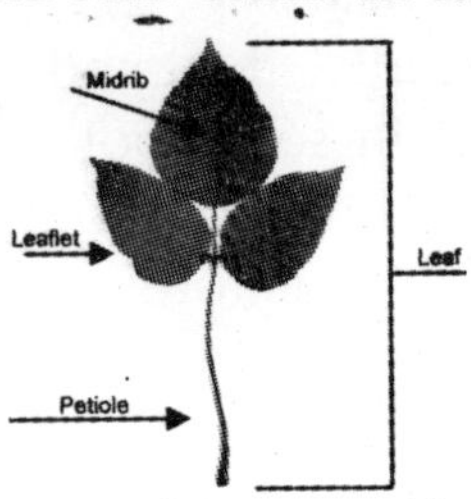

- **lectins**
  plant proteins with a high affinity for specific sugar residues.

- **lectotype**
  one of a series of syntypes that, after publication of the original description, is selected and designated through publication to serve as the type.

- **leggy**
  this is tall and spindly growth, not usual to the growth habit of the plant. It is caused by the lack of light.

- **leghaemoglobin**
  iron-containing, red pigment(s) produced in root nodules during the symbiotic association between rhizobia and leguminous plants. The pigment is similar but not identical to mammalian haemoglobin.

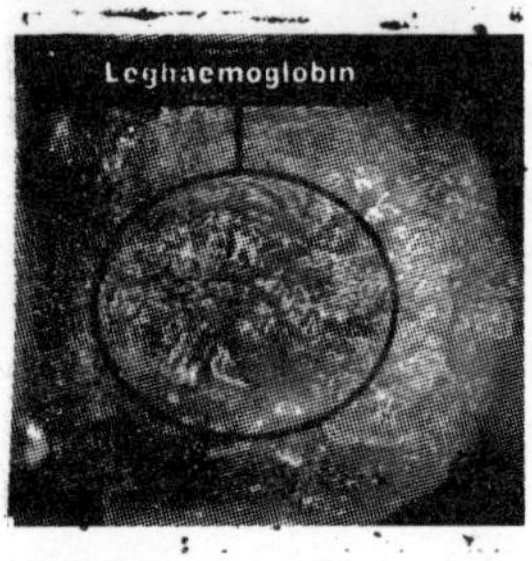

- **legitimate**
  of taxonomic names and epithets: Published in accordance with the Code of Nomenclature.

- **legume**
  a plant whose roots form an association with soil borne bacteria that can capture atmospheric nitrogen.

- **lesion**
  a wound or injury, more specifically, a circumscribed pathological alteration of tissue.

• **library**
in molecular biology: a set of cloned fragments together representing the entire genome.

• **lichen**
a thallus consisting of an alga and fungus intermixed and living in a symbiotic relationship.

• **life cycle**
the complete succession of changes undergone by an organism during its life. A new cycle occurs when an identical succession of changes is initiated.

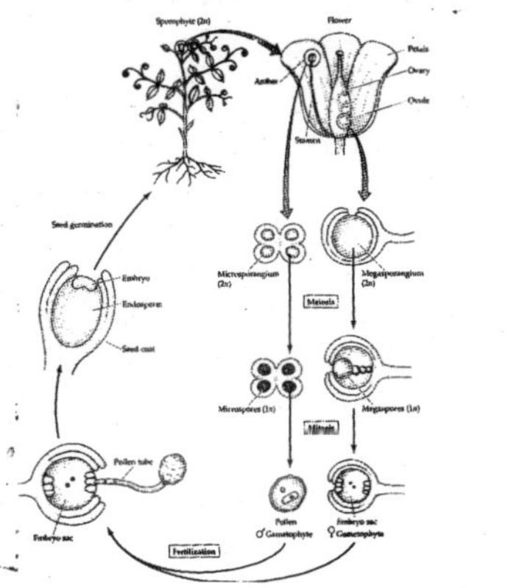

• **ligand**
molecule, ion, or group bound to the central atom in a chelate or a coordination compound.

• **lignify**
to make woody.

• **limbing up**
pruning off the lower limbs of trees, usually for ease of walking underneath or admitting sunlight.

• **lime, agricultural**
soil amendment containing calcium carbonate, magnesium carbonate or other materials to neutralise soil acidity and furnish calcium or magnesium or both for plant growth.

• **limiting factor**
an environmental variable (or, less often, a plant trait) found at a level that restricts the performance of the organism.

• **line variety**
one or more lines of self- or cross-fertilising plants and single-line facultative apomicts, having largely the same genetic background, that are similar in essential and distinctive characteristics and are

maintained or reproduced by controlled self- or sib-fertilisation or line crossing of the plants (for self- or cross-fertilising plants) or close generation control (for single-line facultative apomicts). Usage: The sameness of genetic background is defined as a theoretical coefficient of parentage more than 0.87 and 95 apomixis for the single-line facultative apomicts, except that where it is not possible to achieve 95% apomixis, single-line facultative apomicts with a level of apomixis as low as 80% may be classed as line varieties, even though the variant plants present may differ in morphological characteristics.

- **linear**
  narrow and comparatively long, with parallel margins.

- **lint index**
  weight in grams of fibre from 100 seed.

- **lint quality**
  an estimate of the fitness for use of cotton fibre, based on measurement of various properties of a sample.

- **lip**
  a petal, usually of quite different shape and size to the others, normally at the bottom of the flower, or apparently so, and often, especially in orchids, of complicated structure.

- **LPS**
  lipopolysaccharide. Complex lipid structure containing unusual sugars and fatty acids found in many Gram-negative bacteria.

- **liquid chromatography**
  an analytical method based on the separation of the components of a mixture in solution by selective absorption. All systems contain a moving solvent, a means of producing solvent motion (gravity or a pump), a means of sample introduction, a fractioning column, and a detector. Innovations in functional systems provide the analytical capability for operating in three separation modes:

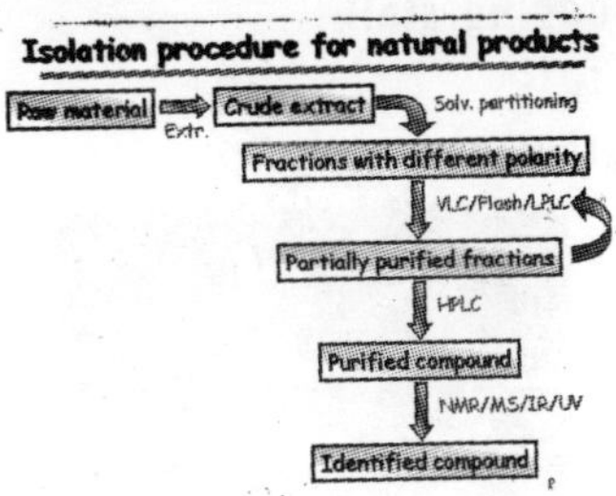

1. liquid-solid partition, in which separations depend on relative solubility of sample components in two immiscible solvents (one of which is usually water).
2. liquid-solid absorption, where the differences in polarities of sample components and their relative absorption on an active surface determine the degree of separation.
3. molecular size separations, which depend on the effective molecular size of sample components in solution.

- **lithophyte**

  a plant that grows on stone using it for anchorage, drawing food and moisture from the air and from humus collected in the crevices of the stone. An 'air-plant.' Orchids generally are found growing one of three ways: lithophytic, epiphytes (the majority grow in this manner), or terrestrials.

- **lithotroph**

  organism that uses an inorganic substrate such as ammonia or hydrogen as an electron donor in energy metabolism. There are two types of lithotrophs: chaemolithotroph and photolithotroph.

- **litter**

  surface layer of the forest floor consisting of freshly fallen leaves, needles, twigs, stems, bark, and fruits.

- **loam**

  good quality soil. Adequate supplies of clay, sand, and fibre must be present. Crumbly to the touch. Ideal for most gardening.

• **lobe**
a part of a segment that represents a division to about the middle.

• **local infection**
an infection affecting a limited part of a plant.

• **local lesion**
a localised spot produced on a leaf upon mechanical inoculation with a virus.

• **locular**
containing chambers or hollows.

• **locule**
a cavity.

• **loculoascomycete**
a member of the Loculoascomycetes.

• **loculoascomycetes**
a group of Ascomycotina with bitunicately discharging asci, producing ascospores that are generally septate and borne in unwalled locules (pseudothecia in ascostromatic ascomata with an ascolocular ontogeny.)

• **locus**
site on a chromosome occupied by a particular gene.

• **lodging, stalk**
stalk breakage above ground level.

• **long day plant**
a plant which requires light for a longer period than it would normally receive from daylight in order to induce flowering, e.g., Saintpaulia.

• **lophotrichous**
having a tuft of polar flagella.

• **low-temperature discolouration**
the loss of chlorophyll and associated green colour that occurs in turf grasses under low-temperature stress.

• **lumen**
in nematodes: Triradiate canal or duct of the oesophagus. In fungi: The space bounded by tissue or wall,

as the central cavity of a cell.

- **luminescence**
  production of light.
- **lunate**
  crescent-shaped, half-moon shaped.
- **luteovirus**
  literally 'yellowish'. Member of a group of plant viruses with isometric particles containing one molecule of linear RNA, mainly confined to the phloem, and usually not mechanically transmitted but transmitted in nature by aphids in a circulative manner.
- **luxury uptake**
  the absorption by plants of nutrients in excess of their need for growth. Luxury contents accumulated during early growth may be used for later growth.
- **lyophilisation**
  rapid freezing of a material at low temperature followed by rapid dehydration by sublimation in a high vacuum. A method used to preserve biological specimens or to concentrate macromolecules with little or no loss of activity. Also known as Freeze-Drying.
- **lyrate**
  shaped like a lyre, with an enlarged apical lobe and smaller lower ones.
- **lysis**
  rupture of a cell, resulting in loss of cell contents.
- **lysogeny**
  an association where a prokaryote contains a prophage and the virus genome is replicated in synchrony with the host chromosome.
- **lysosome**
  cell organelle containing digestive enzymes.
- **macrame**
  decoratively knotted rope or cord forming a harness-like structure for hanging pots.

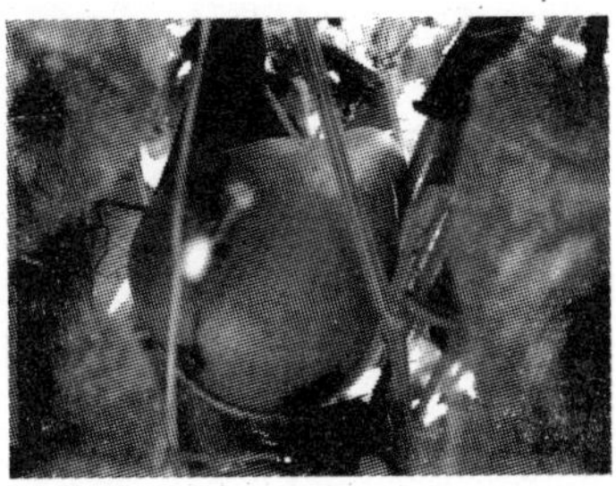

- **macroconidium**
the larger of two types of conidia formed by certain fungi.

- **macrocyclic**
of rusts: the two primary spore stages are present (telial with teliospores and aecial with aeciospores).

- **macrofauna**
soil animals that are more than 1000 micrometers in length (e.g., vertebrates, earthworms, and large arthropods).

- **macromolecule**
large molecule formed from the connection of a number of small molecules.

- **macronutrient**
a substance required in large amounts for growth, usually attaining a concentration of more than 500 mg/kg in mature plants. Usually refers to N, P, K, Ca, Mg, and S.

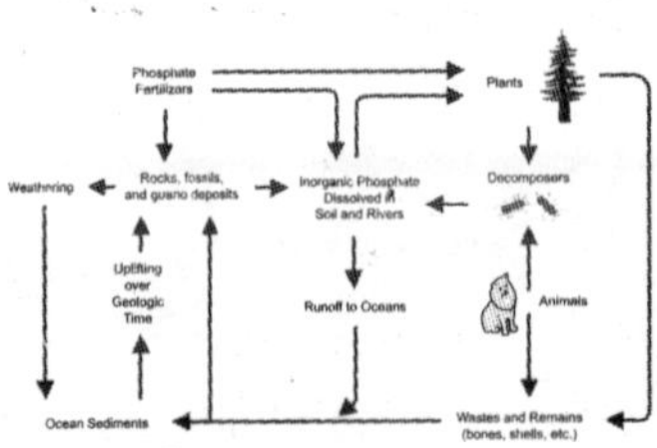

- **macropore**
larger soil pores, generally having a minimum diameter between 30 and 100 micrometers, from which water drains readily by gravity.

- **macroscopic**
visible to the unaided eye.

- **maillard browning reaction**
see **browning**.

- **maincrop**
refers to the crop of vegetables, often potatoes, that produce crops throughout the main growing season.

- **maintenance respiration**
see **respiration**.

- **male-sterile plant**
a plant that produces no functional pollen or is devoid of staminate flowers. See also maintainer line, restorer line.

- **mammillate**
digitate, with nipple-shaped protuberances.

- **manure**
excreta of animals, with or without an admixture of bedding or litter, fresh or at various stages of decomposition or composting. In some countries the term

may denote any fertiliser material.

- **manure tea**
  a liquid fertiliser made by mixing manures with water and filtering out.

- **maquis**
  arid, stony tracts of siliceous soil, covered with shrubs but not trees, such as frequently found in Mediterranean countries.

- **marginal plant**
  plants that will grow on the edges of ponds or lakes and when cultivated will make nice plants around a water garden.

- **marginally hardy**
  close to the limit of hardiness that a plant can withstand. Planting plants that are marginally hardy is risky, because under the most severe conditions for that zone, the plant may not survive without extra protection.

- **marl**
  a chalky clay soil.

- **masked symptoms**
  virus-induced plant symptoms that are absent under some environmental conditions but appear when the host is exposed to certain conditions of light and temperature.

- **mass flow (nutrient)**
  movement of solutes associated with net movement of water.

- **mass planting**
  the planting of one particular flower or many of the same kind somewhat close together to create a dramatic 'look'.

- **mass-flow flux**
  the movement of solutes along with the flow of solvent. The amount moved is the product of solute concentration and the volume of solvent moved.

- **mat**
  thatch, intermixed with soil, that develops between the zone of green vegetation and the original soil surface.

- **matric potential**
  portion of the total soil water potential due to the attractive forces between water and soil solids as represented through adsorption and capillarity.

- **matted-row**
  a system of planting where plants are placed off centre or are cantered on a diagonal.

- **mature fibres**
  in cotton: fibres that, when treated with a sodium hydroxide (NaOH) solution, swell into a non-convoluted and almost rod like shape, have a total wall width equal to or greater than the lumen width.

- **maturity index**
  any mechanical, physiological, chemical, visual or other method used to determine the time of maturity for a plant species.

- **mean length, array**
  in cotton: the average length of all the fibres in the test specimen based on weight-length data.

- **mechanical innoculation**
  of plant viruses, a method of experimentally transmitting the pathogen from plant to plant, juice from diseased plants is rubbed on test-plant leaves that usually have been dusted with carborundum or some other abrasive material.

- **medulla**
  central part of an organ.

- **meiospore**
  a uninucleate, haploid spore arising directly by meiosis.

- **membranaceous**
  thin and more or less translucent.

- **mentum**
  the chin-like protuberance occurring in certain orchid flowers, formed usually by the bases of the lateral sepals with the elongated column-foot.

- **mericlone**
  an exact genetic copy of another plant produced by meristem culture.

- **meristem**
  tissue composed of dividing cells to produce tissues and organs, located in small

amounts within the growth buds and root tips. The growing point of shoots.

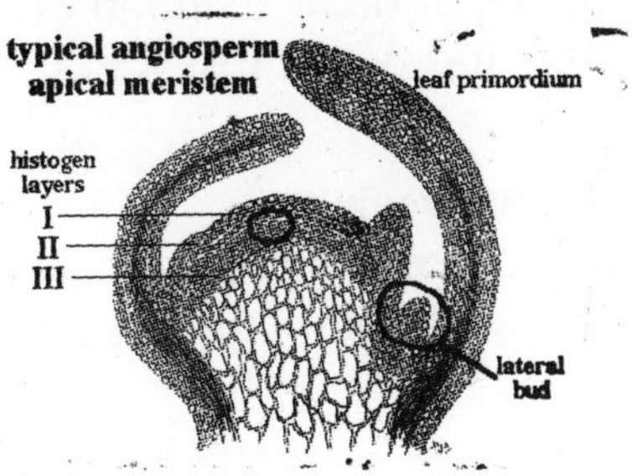

- **meristem culture**
a laboratory technique that involves the taking of the growing meristem tip from within the new growth and culturing the nucleus of cells, in a similar way to germination of orchid seeds artificially.

- **mesochile**
the intermediate or middle part of the lip of orchids when this structure is separated into three distinct parts, as in Stanhopea.

- **mesofauna**
soil animals between 200 to 1000 micrometers in length, including nematodes, oligochaete worms, smaller insect larvae, and small arthropods.

- **mesophile**
organism whose optimum temperature for growth falls in an intermediate range of approximately 15° C to 40° C.

- **mesophyll**
the leaf cells that contain chloroplasts and that are located between the upper and lower epidermis.

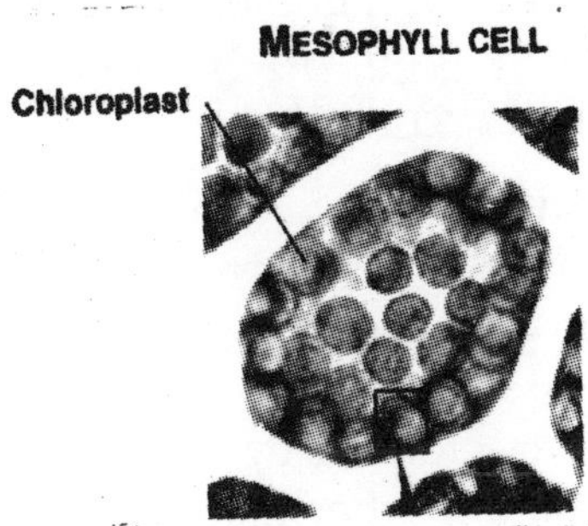

- **messenger RNA**
a chain of specific ribonucleotides that codes for a specific protein, template for the assembly of amino acids into protein. In cells mRNA is transcribed from DNA, but some RNA viruses function directly as mRNA.

- **metabasidium**
the cell in which meiosis occurs in members of the Basidiomycotina. (pl. metabasidia).

- **metabolism**
all biochemical reactions in a cell, both anabolic and catabolic.

- **metacorpus**
the swollen posterior portion of the corpus, sometimes called the median bulb.

- **metamorphosis**
the changes of form insects go through in their life cycle from egg to immature stages to adult.

- **metaplasia**
changed condition of a structure or organ, hyperplastic class of symptoms characterised by overdevelopment other than that due to hypertrophy or hyperplasia e.g., abnormal starch accumulation, virescence, etc.

- **methanogenesis**
biological production of methane.

- **methanogenic bacterium (methanogen)**
methane-producing prokaryote, member of the Archaea.

- **methanotroph**
organism capable of oxidising methane.

- **micro-propagation**
also called propagating by tissue culture. Taking cell slices of plant material and starting growth on culture dishes.

- **micro aerophile**
organism that requires a low concentration of oxygen for growth. Sometimes indicates an organism that will carry out its metabolic activities under aerobic conditions but will grow much better under anaerobic conditions.

- **micro aggregate**
clustering of clay packets stabilised by organic matter and precipitated inorganic materials.

- **microbial biomass**
total mass of micro-organism alive in a given volume or mass of soil.

- **microbiology**
study of micro-organisms.

- **microclimate**
the warmth and humidity of the air in close proximity to a plant. It may differ significantly from the general climate of the room.

- **microconidium**
  1. a spermatium.
  2. a small conidium. (pl. microconidia)
- **microcosm**
  a community or other unit that is representative of a larger unity.
- **microcutting**
  1. a plant produced by micropropagation.
  2. a modern technique using tiny pieces of the parent plant on a sterile nutrient jelly.
- **microcyclic**
  a life cycle in the rusts where one or more of the main spore stages, usually the aecia, is absent.
- **microenvironment**
  immediate physical and chemical surroundings of a micro-organism.
- **microfauna**
  protozoa, nematodes and arthropods generally less than 200 micrometers long.
- **microflora**
  bacteria (including actinomycetes), fungi, algae, and viruses.
- **microhabitat**
  clusters of micro aggregates with associated water within which microbes function. May be composed of several micro sites (e.g., aerobic and anaerobic).
- **micrometer**
  one-millionth of a meter, or 10^{-6} meter, the unit usually used for measuring micro organisms.
- **micronutrient**
  chemical element necessary for growth found in small amounts, usually less than 100 mg kg^{-1} in a plant. These elements consist of B, Cl, Cu, Fe, Mn, Mo, and Zn.
- **micro-organism (microbe)**
  living organisms-too small to be seen with the naked eye (less than 0.1 mm), include bacteria, fungi, protozoan, microscopic algae, and viruses.
- **micro-pore**
  relatively small soil pore, generally found within structural aggregates and having a diameter less than 30 micrometers.

- **micro-pyle**
  in nematodes: A minute opening in the membrane of an egg through which the spermatozoa enter.

- **microsclerotium**
  a small clump of dark-coloured, more or less thick-walled cells, each of which is viable, produced in culture and rarely in the xylem of host plants (pl. microsclerotia).

- **microscopic**
  very small, seen only with the aid of a microscope.

- **microsite**
  small volume of soil where biological or chemical processes differ from those of the soil as a whole, such as an anaerobic microsite of a soil aggregate or the surface of decaying organic residues.

- **middlings**
  a by-product of flour milling that contains varying proportions of endosperm, bran, and germ.

- **mildew**
  several different types of fungi. Two popular types are downy and powdery. It leaves a white coating on the leaves. Common to crepe myrtle, zinnias, grapes, and roses even bee balms. It shows up in cool, wet weather. Drip irrigation can prevent some occurrence. Plant disease resistant varieties.

- **mineralisation**
  conversion of an element from an organic form to an inorganic state as a result of microbial decomposition.

- **minimal medium**
  a type of culture medium lacking specific growth factors. It does not support the growth of some or all auxotrophic strains of a given organism but permits the growth of prototrophic strains.

- **minimum tillage**
  1. minimal soil manipulation in combination with chemicals for adequate seedbed preparation and vegetation control.
  2. minimal soil manipulation in combination with chemicals and residue incorporation for minimum moisture loss, reducing energy input and labour re-

quirement. Synonym: conservation tillage.

- **mist propagation**
  the ideal method of propagation in a green house or place with good circulation, using automatic misters.

- **miticide**
  a pesticide that kills mites.

- **mitochondrion (plural, mitochondria)**
  eukaryotic organelle responsible for processes of respiration and oxidative phosphorylation.

- **mitosis**
  a sequence of cellular events that culminates in the division of a eukaryotic nucleus into two genetically similar or identical nuclei whose ploidy is the same as that of the parent nucleus. Mitosis occurs during asexual cell division. pl. mitoses.

- **mitospore**
  a uninucleate, haploid or diploid spore arising by mitosis.

- **mitosporic fungi**
  see **deuteromycotina**.

- **mixed border**
  some of this, some of that in a border. Shrubs and woody plants that are still in the bed when the annuals and perennials are gone for the season.

- **mixotroph**
  organism able to assimilate organic compounds as carbon sources while using inorganic compounds as electron donors.

- **mixture, seed**
  1. seed of more than one kind or cultivar, each present as more than 5 per cent by weight of the whole. 2. a combination of seed of two or more species.

- **MLO**
  Mycoplasma-Like Organism.

- **mobilisation**
  chemical change of stored materials (see **reserves**) that enables translocation of the compounds or their use for synthesis or energy. Usage: The term remobilisation is sometimes used to denote mobilisation of structural materials (e.g., proteins during senescence).

- **modulus**
the ratio of change in stress to change in strain in the initial straight-line portion of a stress-strain curve following the removal of any crimp.

- **modulus, secant**
the ratio of change in stress to change in strain between two points on a stress-strain diagram, particularly the points of zero stress and breaking stress.

- **moist but not wet**
plants that grow well in moist conditions but not entirely wet. Whether it be in the ground or in a container.

- **moisture equilibrium**
1. in seed: equilibrium of the moisture content of the seed and the relative humidity of the surrounding air.
2. more generally, the condition reached by a sample when it no longer takes up moisture from, or gives up moisture to, the surrounding atmosphere.

- **mould**
a downy fungal growth on a substratum, usually consisting of mycelium of a Hyphomycete or a Zygomycete.

- **molecule**
result of two or more atoms combining by chemical bonding.

- **mollicute**
a proposed trivial name for any member of the class Mollicutes. The use of this name could permit mycoplasmas' to be used specifically for members of the genus Mycoplasma.

- **mollusc**
invertebrate animals with soft, unsegmented bodies, such as clams and snails, usually enclosed in a calcium shell.

- **molt**
to shed the exoskeleton to accommodate growth.

- **monandrous**
with one stamen.

- **monilia**
reported to be allergenic. This fungus produces soft rot of tree fruits. Other members produce a red bread mould. It is infrequently involved in corneal eye infections.

- **monocarpic**
plants that only flower and fruit once in their lives.

- **monoclonal antibody**
antibody produced from a single clone of cells. This antibody has uniform structure and specificity.

- **monocot**
or monocotyledon, flowering plants that have embryos with only one cotyledon.

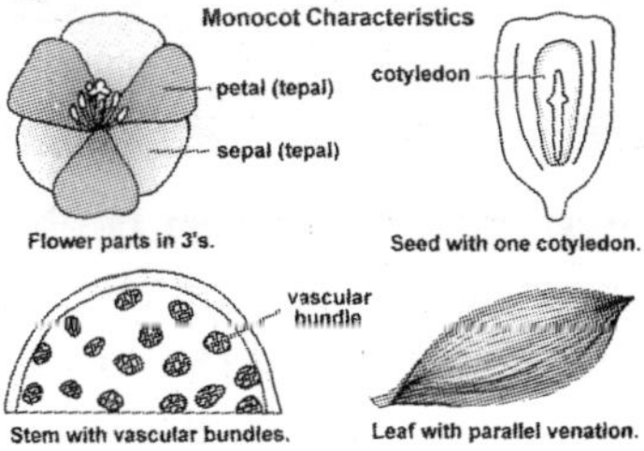

- **monocotyledon**
with a single cotyledon or seed-leaf.

- **monoculture**
the use of one crop or family of crops either continuously, year after year, or alternated with fallow.

- **monocyclic**
of a disease or pathogen: producing one generation of innoculum and one cycle of infection during a single growing season.

- **monodelphic**
of nematodes: Possessing one genital tube or ovary.

- **monoecious**
with the male (staminate) flowers and the female (pistillate) flowers borne in separate inflorescences but on the same plant.

- **monogenic**
of nematodes: producing offspring of only one sex.

- **monogenic resistance**
resistance determined by a single gene.

- **monokaryon**
fungal hypha in which compartments contain one nucleus.

- **monokaryotic**
  having one nucleus per cell.
- **monopodium**
  orchids that grow primarily upwards, producing new growth at the top of the plant from the location of the previous growth. Leaves are produced alternately on either side of the central stem as it grows. Orchids with a monopodial growth pattern are less common than those with a sympodial growth pattern.
- **monostand**
  a turf grass community composed of only one cultivar.
- **monotype**
  the sole species of a newly proposed genus.
- **moon gardening**
  a very romantic night garden created by using white night flowering plants or light variegated leaf varieties of plants. Really a pretty effect.
- **morphologic**
  pertaining to form
- **mosaic**
  a common symptom induced in leaves by many plant virus infections in which there is a pattern of dark green, light green and sometimes chlorotic areas. This pattern is often associated with the distribution of veins in the leaf. In monocotyledonous leaves it shows as stripes.
- **moss**
  small, leafy plants that do not produce flowers or seeds. They grow in moist, shaded areas where fertility is low.

- **Most Probable Number (MPN)**
  method for estimating microbial numbers in soil based on extinction dilutions.
- **motes**
  small black specks seen in cotton yarn or cloth.
- **motility**
  movement of a cell under its own power.

• **mottle**
a diffuse form of the mosaic symptom in plant leaves in which the dark and light green are less sharply defined. This term is frequently used interchangeably with mosaic.

• **mouth**
the open end of a bell shaped or tubular flower.

• **mowing frequency**
1. the number of mowing per unit of time (usually, per week).
2. the interval in days between one mowing and the next.

• **mowing pattern**
the orientations of travel while mowing turf. Patterns may regularly be changed to distribute wear and compaction, to aid in grain control, and to create visually aesthetic striping effects.

• **mowing strip**
sometimes known as an edging strip. The six to eight inches of space between the lawn and the flower or vegetable bed. The flat area decoratively made with cement or brick.

• **mucigel**
gelatinous material at the surface of roots grown in normal non-sterile soil. It includes natural and modified plant mucilages, bacterial cells, and their metabolic products (e.g., capsules and slimes), and colloidal mineral and organic matter from the soil.

• **mucilage**
gelatinous secretions and exudates produced by plant roots and many micro-organisms.

• **mucor**
often found in soil, dead plant material, horse dung, fruits and fruit juice. It is also found in leather, meat, dairy products, animal hair and jute. A Zygomycetes fungus which may be allergenic (skin and bronchial tests). This organism and other Zygomycetes will grow rapidly on most fungal media.

May cause mucorosis in immune compromised individuals. The sites of infection are the lung, nasal sinus, brain, eye and skin. Infection may have multiple sites.

- **mucro**
  in nematodes: A stiff or sharp point abruptly terminating an organ.

- **mucronate**
  in nematodes: Ending in a sharp point.

- **mulch**
  a protective covering that is spread on the ground around plants to inhibit evaporation and weed growth, control soil temperature, enrich the soil, or prevent the dispersal of pathogens. It may be organic material such as leaves, peat, or wood chips, or inorganic material such as plastic sheeting.

- **multicolour**
  a flower with petals which bear at least three distinctly different colours.

- **multicomponent virus**
  a virus in which the genome needed for full infection is divided between two or more particles e.g., cowpea mosaic virus, brome mosaic virus, cucumber mosaic virus.

- **multiline**
  a plant cultivar made up of a mixture of isolines differing in their genes for resistance to a particular pathogen.

- **multiseptate**
  having more than one septum.

- **multispecies grazing**
  usage: Not an acceptable term: 'multi' refers to 'many', while this grazing method most often employs only two animal species.

- **mummy**
  a dried, shrivelled fruit colonised by a fungus.

- **municipal solid waste**
  combined consumer and commercial waste gener-

ated within a defined geographic area.

- **muriform**
  having bricklike cells in a wall with both longitudinal and transverse septa.

- **mushroom**
  a fleshy fruiting body of a fungus, especially of a basidiomycete of the family Agaricaceae.

- **mutagen**
  substance that causes the mutation of genes.

- **mutant**
  organism, population, gene, or chromosome that differs from the corresponding wild type by one or more base pairs.

- **mutation**
  a stable, heritable change in the nucleotide sequence of a genetic nucleic acid (DNA, or RNA)in viruses, viroids, etc typically resulting in the generation of a new allele and a new phenotype.

- **mutualism**
  interaction between organisms where both organisms benefit from the association.

- **mycelium**
  a mass of hyphae, often used to denote all hyphae comprising a thallus. (pl. mycelia).

- **mycology**
  the study of fungi

- **mycoparasite**
  a fungus parasitic on other fungi.

- **mycophagous**
  organisms that consume fungi, such as mycophagous nematodes.

- **mycoplasma**
  1. group of bacteria without a cell walls that do not revert to walled forms. Phylogenetically related to clostridia.
  2. a genus of cell wall-less, sterol-requiring, catalase-negative bacteria (family Mycoplasmatace) occurring as parasites and pathogens.

- **mycorrhiza**
  a specialised root structure resulting from a symbiotic relationship between a fungus and a higher plant.

- **mycorrhizosphere**
  unique microbial community that forms around a mycorrhiza.

- **mycosis**
  disease in animals caused by a fungus. (pl. mycoses).

- **mycotoxicosis**
  any disease of man or animals resulting from the ingestion of mycotoxins.

- **mycotoxin**
  a toxin produced by a fungus. The term is usually reserved for fungal metabolites that are toxic to man and/or animals and are produced by moulds growing on foodstuffs (e.g., aflatoxins, ergot alkaloids.)

- **mycotrophic**
  refers to green plants having mycorrhizae.

- **mycovirus**
  a virus that replicates in cells of fungi.

- **myrmecophilous**
  ant-loving, inhabited by ants.

- **myxamoeba**
  a naked cell capable of amoeboid movement, characteristic of the vegetative phase of myxomycetes and such Plasmodiophoromycetes as *plasmodiophora brassicae.*

- **myxomycota**
  the slime moulds, a class of fungi characterised by amoeboid vegetative protoplasts, plasmodia, and by brightly coloured spore bearing capillitia.

- **n-acetylglucosamine and n-acetylmuramic acid**
  sugar derivatives in the peptidoglycan layer of bacterial cell walls.

- **nanopore**
  soil pore having dimensions measured in nanometers. Materials encased in nanopores are beyond the reach of micro-organisms and enzymes.

- **native**
  this refers to a plant that grows in the same habitat in which they originated. These plants can be native to a continent, state, or region.

- **native species**
  a species indigenous to an area, i.e., not introduced from another environment or area.

- **naturalised**
  plants that will behave like native plants in a given geological region. Bulbs naturalise nicely and lend themselves to a blooming statement.

- **naturalised plants**
  introduced species that have become established in a region.

- **near-isogenic lines**
  two distinct composites of F3 lines from a single cross, one consisting of lines homozygous recessive and the other consisting of lines homozygous dominant for specific genes. That is, the paired composite lines have the same genetic background, differing only in being homozygous dominant vs. recessive for the specific genes.

- **necrosis**
  damage of living tissues because of infection or injury.

- **necrotic**
  dead.

- **necrotroph**
  1. an organism that kills part or all of another organism before deriving nutrients from it (usually applied to plant pathogens.)
  2. an organism that derives nutrients from dead plant or animal tissues, whether or not it is responsible for the death of those tissues.

- **nectar**
  a sugar and water substance secreted by flowers, to attract pollinators like bees and hummingbirds searching for food.

- **nectariferous**
  having nectar

- **neem**
  a botanical insecticide that is non-toxic. It is derived

from the neem tree *(azaderachta indica).*

- **nematicide**
a chemical compound or physical agent that kills nematodes.

- **nematode**
more or less elongate, spindle-shaped, worm-like animals ranging in size from less than a millimetre to several meters in length, living as saprophytes in soil or water or as parasites of plants or animals.

- **neotype**
a specimen selected as the type subsequent to the original description in cases in which the primary types are definitely known to have been destroyed.

- **nepovirus**
siglum of nematode polyhedral virus. Member of a group of multicomponent plant viruses with two isometric particles containing two species of linear RNA, transmitted mechanically and by soil-inhabiting nematodes.

- **nerve ring**
the centre of the nervous system of nematodes that encircles the oesophagus, composed largely of nerve fibres and associated ganglia.

- **neutral**
neither acid nor alkaline with a pH of around 7.

- **neutral-detergent fibre**
neutral-detergent insoluble residue, primarily cell-wall constituents.

- **neutralism**
lack of interaction between two organisms in the same habitat.

- **new wood**
a term often used in reference to propagation. That part of the stems and branches that have grown during the current season. Some plants will propagate better on new wood as opposed to old wood (previous season's growth).

• **niche**
functional role of a given organism within its habitat.

• **nick**
the state or condition in which two parents for producing hybrid seed mate well to produce superior yields or progeny.

• **nigrospora**
reported to be allergenic.

• **nitrate reduction (biological)**
process whereby nitrate is reduced by plants and micro organisms to ammonium for cell synthesis (nitrate assimilation, assimilatory nitrate reduction) or to various lower oxidation states (N_2, N_2O, NO) by bacteria using nitrate as the terminal electron acceptor in anaerobic respiration.

• **nitrate toxicity**
a variety of conditions in animals, resulting from ingestion of feed high in nitrate, the toxicity actually results when nitrate (NO_3) is reduced to nitrite (NO_2) in the rumen.

• **nitrification**
biological oxidation of ammonium to nitrite and nitrate, or a biologically induced increase in the oxidation state of nitrogen.

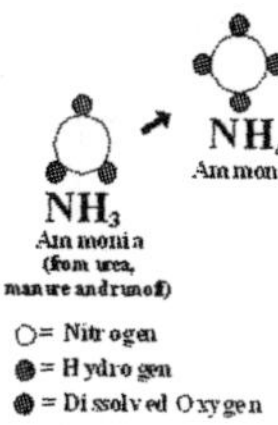

• **nitrifying bacteria**
chemolithotrophs capable of carrying out the transformations from NH_3 to NO_2 or NO_2 to NO_3.

• **nitrogen cycle**
sequence of biochemical changes wherein nitrogen is used by a living organism, transformed upon the death and decomposition of the organism, and converted ultimately to its original state of oxidation.

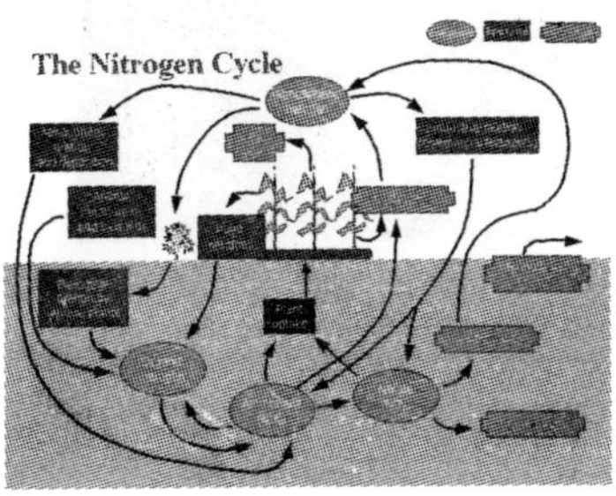

- **nitrogen fixation**
  the capture and conversion of atmospheric nitrogen gas into nitrogen compounds, stored in the soil that can be used by plants.

- **nitrogenase**
  specific enzyme system required for biological N_2 fixation.

- **nitrogen-free extract**
  the unanalysed portion of a plant, consisting mostly of carbohydrates, that remains after the protein, ash, crude fibre, ether extracts, and moisture content have been determined.

- **nocturnal**
  of the night, used in reference to flowers which open after dark.

- **nodal cutting**
  a cutting taken from a piece of stem cut just below a node.

- **node**
  the point on a stem where a leaf or bud is attached, the place which is used for propagation.

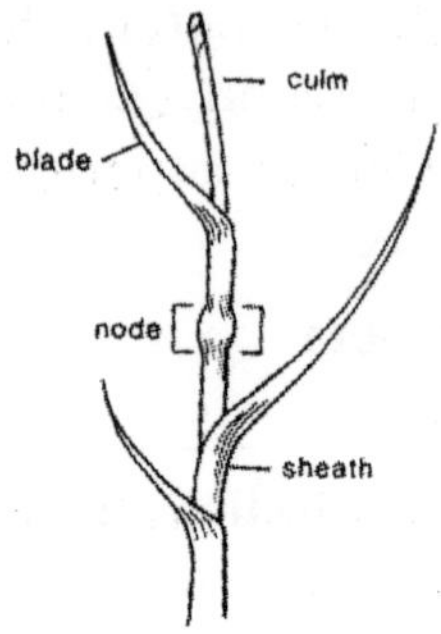

- **nodules**
  swellings on the roots of legumes where nitrogen-fixing bacteria live.

- **nomenclatural synonym**
  any one of two or more different names based on the same plant specimen.

- **nomenclatural type**
  that plant with which the name is permanently associated, not necessarily the most typical or representative of the taxon.

- **nomenclature**
  a system of names, or naming, as applied to the subjects or study in any art or science, especially in botany and zoology.

• **noncirculative transmission**
virus transmission characterised by a very short period of acquisition of the virus by a vector (e.g., an aphid), no latent period before the vector can transmit the virus, and a short period of retention by the vector after acquisition

• **non-nutritive fibre**
that portion of fibre in a feed that is not digestible and hence is of no nutritive value.

• **nonpolar**
possessing hydrophobic (water repelling) characteristics and not easily dissolved in water.

• **non-resupinate**
orchid flowers normally position the lip at the bottom just above the column. Some genera, however, such as Cycnoches, Malaxis, and Nephelaphyllum position the lip uppermost with the column below making the flower appear to be up-side-down.

• **non-selective pesticide**
a poison that kills a wide variety of pest species.

• **non-septate**
lacking cross-walls.

• **non-vascular**
refers to organisms with no tissues or vessels to carry water, minerals, etc., such as mosses, fungi, algae, lichens, etc.

• **northern blot**
hybridisation of single-stranded nucleic acid (DNA or RNA) to RNA fragments immobilised on a filter.

• **nostoc**
a blue-green alga with filaments formed of chains of cells.

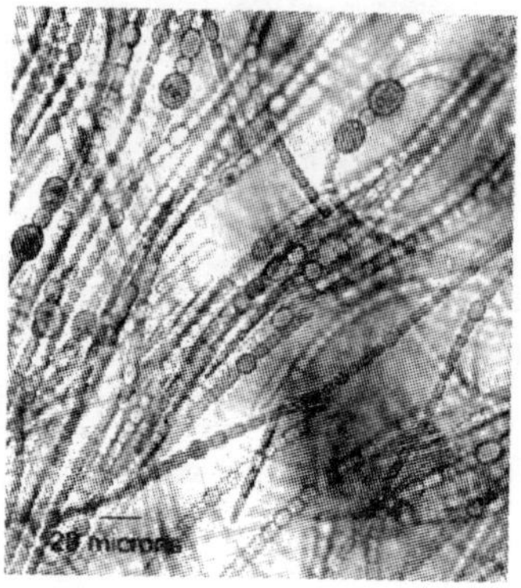

• **notch**
a V-shaped indentation.

• **nothomorph**
a term used in taxonomy to designate different hybrid forms derived from the same parent species, ranking essentially as a variety.

• **no-till**
a method of planting crops that involves no seedbed preparation other than opening small areas in the soil for placing seed at the intended depth. There is generally no cultivation during crop production, instead, chemicals are used for vegetation control.

• **novirame**
a flowering, later fruiting, shoot arising from a primocane.

• **noxious weeds**
plants which may cause harm to collectors, such as poison ivy, Toxicodendron radicans, or invasive exotics

or parasites and their host plants which may harm the ecosystem or agriculture of an area.

• **nucellus**
the central body of the ovule that encloses the female gametophyte, equivalent to the megasporangium.

• **nuclease**
any enzyme that can cleave the sugar-phosphate backbone of a nucleic acid.

• **nucleic acid**
polymer of nucleotides.

- **nucleoid**
aggregated mass of DNA that makes up the chromosome of prokaryotic cells.

- **nucleolus**
a special body in the nucleus.

- **nucleophilic compound**
chemical that attracks or is drawn to electron-deficient regions in other chemicals, reducing agents act as nucleophilic compounds.

- **nucleoprotein**
a compound of nucleic acid and protein.

- **nucleoside**
nucleotide without the phosphate group.

- **nucleotide**
monomeric unit of nucleic acid, consisting of a sugar (pentose), a phosphate, and a nitrogenous base.

- **nucleus**
1. the germ cell of the ovule, which by fertilisation becomes the seed, the kernel of a seed.
2. the part of the protoplasm that contains the chromosomes. (pl. nuclei)

- **nullipores**
calcareous red seaweed.

- **nurse grass**
a fast-growing temporary grass included in most grass seed mixtures to give rapid coverage and protect permanent grasses from soil erosion, sunscald, etc.

- **nurse tree**
a fast-growing tree planted with, or left in place near, a slow-growing, permanent tree in order to provide shade and wind protection for the permanent tree until it has reached useful size.

- **nurseryman**
one who is state certified to practice growing techniques of plant material making the proper selection for specific needs.

- **nut**
a dry, hard indehiscent 1-celled and 1-seeded fruit, though usually resulting from a compound ovary.

• **nutant**
nodding, usually referring to the entire inflorescence rather than a single flower, cernuous.

• **nutlet**
one of several small, nut-like parts of a compound fruit, a diminutive nut.

• **nutrient**
substance taken by a cell from its environment and used in catabolic or anabolic reactions.

• **nutrient broth**
a liquid basal medium.

• **nutrient, animal**
food constituent or group of food constituents of the same general chemical composition required for support of animal life.

• **nutritive value index**
daily digestible amount of forage per unit of metabolic body size relative to a standard forage.

• **nymph**
a stage or series of size changes between egg and adult in the life cycle of insects that go through incomplete or simple metamorphosis.

• **OC**
On Centre, refers to the spacing on landscape plans of materials to be planted.

• **ob**
a prefix meaning inverted, e.g., obcordate describes a heart-shaped leaf attached at the point rather than at the cleft.

• **obcompressed**
compressed dorso-ventrally instead of laterally.

• **obconically**
inversely conical, having the attachment at the apex.

• **obcordate**
inverted and heart-shaped, e.g., a heart-shaped leaf with the pointed end toward the stem.

- **obcuneate**
like cuneate, but with the point of attachment at the broad end.

- **obelisk**
an upright four-sided usually monolithic pillar that gradually tapers as it rises and terminates in a pyramid, often inscribed with words or designs.

- **obhastate**
like hastate, but with the triangular lobes at the tip.

- **oblanceolate**
lanceolate with the broadest part toward the apex.

- **oblate**
shaped like a sphere which is depressed at the poles.

- **obligate**
1. adjective referring to an environmental factor (for example, oxygen) that is always required for growth.
2. organism that can grow and reproduce only by obtaining carbon and other nutrients from a living host, such as obligate symbiont.

- **obligate anaerobe**
an organism that can grow only under anaerobic conditions.

- **obligate parasite**
an organism that is incapable of living as a saprophyte and must live as a parasite.

- **oblique**
unequal-sided or slanting.

- **obloid**
having an oblong shape but with the cross section circular and the ends rounded.

- **oblong**
of spores: longer than broad (about twice as long or somewhat less), with sides nearly parallel and with ends more or less flattened.

- **obovate**
inverted ovate, egg-shaped, with the broadest part above, or away from the stem.

• **obovoid**
egg-shaped, with the narrow end outward.

• **obpyramidal**
shallow triangular with the attachment at the point rather than the middle of the flat side.

• **obpyriform**
inversely pear-shaped, with the attachment at the narrow end, turbinate.

• **obsolescent**
becoming rudimentary, refers to an organ no longer functional and reduced to vestigial remains.

• **obtuse**
of pileus, cystidia, spores rounded or blunt.

• **ocellated**
having spots.

• **ochrea**
the sheath around the base of the seta, terminating the vaginula.

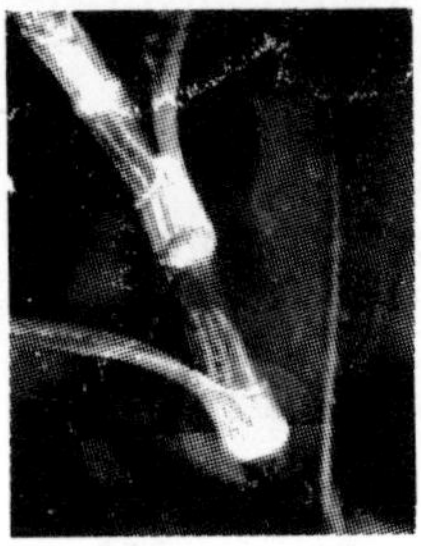

• **ocrea**
a legging-shaped or tubular stipule.

• **ocreate**
having sheathing stipules.

• **octoploid**
having eight sets of chromosomes.

• **odd pinnate**
a pinnate leaf with a terminal leaflet, hence having an odd number of leaflets.

• **odontostylet**
see **stylet**. Synonymous with onchiostylet.

- **odouriferous**
producing an odour, often fragrant.

- **oedema**
intumescence or blister formation because of an increase in intercellular water. Also known as Edema.

- **officinal**
of the shops, a plant used in medicine or the arts.

- **offset**
a young plantlet which appears on a mature plant. An offset can generally be detached and used for propagation as in spider plants or walking iris.

- **offshore wind**
blowing from the land across the water, usually warmer than an onshore wind, and with lower waves.

- **off-type**
1. plants or seeds that do not conform to the characteristics of a variety as described by the breeder.
2. plants or seeds that are not necessarily of any variety. Plants or seeds resulting from cross-pollination by other kinds or varieties, and uncontrolled self-pollination during production of hybrid seed. See also **roguing**.

- **old garden rose**
a hybrid rose which has been in cultivation since before 1867.

- **oligo**
a prefix denoting few, as oligopetalous means with few petals.

- **oligonucleotide**
short nucleic acid chain, either obtained from an organism or synthesised chemically.

- **oligotroph**
micro-organism specifically adapted to grow under low nutrient supply. Thought to subsist on the more resistant soil organic matter and be little affected by the addition of fresh organic ma-

terials. Sometimes a synonym for autochthonous.

- **once-compound**
a compound leaf with a single set of undivided leaflets.

- **onchiostylet**
in nematodes: a stylet developed from a special cell in the anterior part of the oesophagus from which it moves into place during each moult.

- **one year bedded**
refers to plants which have been treated to line-out, planted thickly, and grown for one year following propagation.

- **onshore wind**
blowing from the water to the shore, building large waves as it uses its full fetch. This is usually cooler than an offshore wind.

- **ontogeny**
the course of development of a single organism.

- **oocyst**
synonymous with oogonium.

- **oocyte**
a cell from which an egg develops.

- **oogamy**
a type of heterogamy in which plasmogamy takes place between a large nonmotile egg and a small motile male gamete or cytoplasm from an antheridium.

- **oogonium**
specialised sexual structure formed as a female gametangium by fungus like organisms in the phylum Oomycota.

- **oomycetes**
a class of aquatic and terrestrial fungi (subdivision Mastigomycotina) that typically produce oogonia and zoosporangia in which form zoospores having one anteriorly-directed tinsel flagellum and one posteriorly-directed whiplash flagellum.

- **oosphere**
a large, naked, nonmotile, usually spherical, female gamete.

- **oospore**
a thick-walled spore that develops from an oosphere

through plasmogamy or parthenogenesis.

- **opaque**
an endosperm mutant of maize associated with suppressed prolamine production in the endosperm which results in increased lysine content of the protein fraction.

- **open pollinated**
any plant that has been pollinated in the field. In direct

contrast to hybrid varieties that will grow true to variety (the plants produced will be true to the parents.)

- **open pollination**
pollination by natural mechanisms such as insects, wind, etc., as opposed to selective pollination by a plant breeder.

- **open storage**
storage (e.g., of seed) with free access to normal atmospheric conditions.

- **operculate**
of an ascus or sporangium, opening by a sub-circular apical lid to discharge spores, as in asci of the Pezizales.

- **operculum**
a lid, the upper portion of a circumscissile capsule, which on detachment permits the spores to escape.

- **operon**
cluster of genes whose expression is controlled by a single operator, typical in prokaryotic cells.

- **opposite**
leaf form, where the leaves are arranged in opposite pairs along the stem.

- **orangery**
a many-windowed building used to house potted orange trees during winter.

- **orbiculate**
circular in outline.

- **orchard**
a planting of fruit or nut trees.

- **orchidaceae**
pertaining to a large family of perennial epiphytic or terrestrial plants, orchid family.

- **orchidaceous**
  orchid family, usually having showy flowers with corolla of three petals, one labellum or lip differs greatly from others and often spurred.

- **orchidist**
  one who collects or is interested in orchids horticulturally.

- **orchidologist**
  a botanist who specialises in the technical sturdy of orchids.

- **order**
  a category of taxonomic classification ranking above the family and below the class.

- **organelle**
  a membrane-bound body analogous to an organ and found in the cytoplasm of the cell that performs specific cellular functions.

- **organic**
  fertilisers and chemicals that have been obtained from a source which is or has been alive. Also the general term used for a type of gardening using no chemical or synthetic fertilisers or pesticides.

- **organic soil**
  soil that contains a high percentage (more than 200 g/ kg^{-1}, or more than 120-180 g/ kg^{-1} if saturated with water) of organic carbon.

- **organogenesis**
  the process of initiation and development of callus into one distinct organ (e.g., roots or shoots). Initiation is removed in time and space from the initiation of other organs. Such organs may be adventitious or de novo in origin.

- **organotroph**
  organism that obtains reducing equivalents (stored electrons) from organic substrates.

- **ornamental**
  a plant that is grown strictly for its foliage or flower rather than for food or any

other economic use (e.g., saucer magnolia.).

- **ornamental tree**
generally, a small tree as opposed to a tall, or shade, tree. Often they are flowering trees used as under story plantings, or massed in the open for colour and texture.

Koelreuteria paniculata
Golden Rain Tree

- **ortet**
the original plant from which the members of a clone have descended.

- **orthodox seed**
seed that can be dried to moisture levels between 4 and 6 percent and stored without spoiling.

- **orthotropic**
tending to grow or form along a vertical axis.

- **orthotropous**
describes an ovule or seed that is erect, with the orifice or micropyle at the apex.

- **osmosis**
the diffusion of liquid through a semi-permeable membrane (such as a cell wall) until there is an equal concentration on both sides of the membrane. The process by which water is taken up through the roots of a plant and transported throughout the structure.

- **osmotic potential**
portion of total soil water potential due to the presence of solutes in soil water.

- **osmunda fibre**
the roots of the fern Osmunda regalis, used for making Orchid Compost.

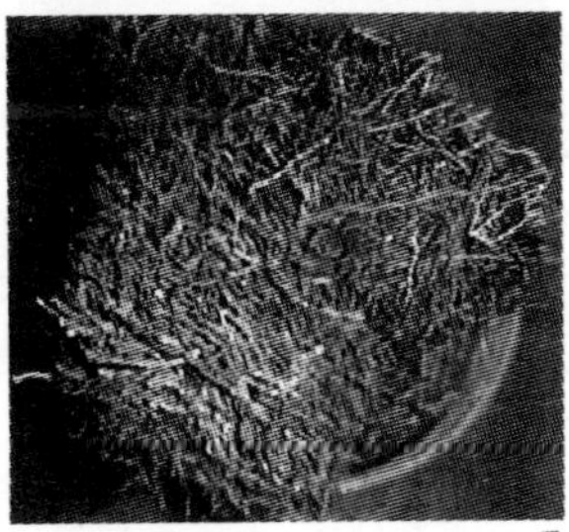

- **osseous**
bony and brittle.

- **ostiole**
1. a neck-like structure in an ascocarp, lined with pariphyses, and terminating in a pore.

2. the opening of a pycnidium.

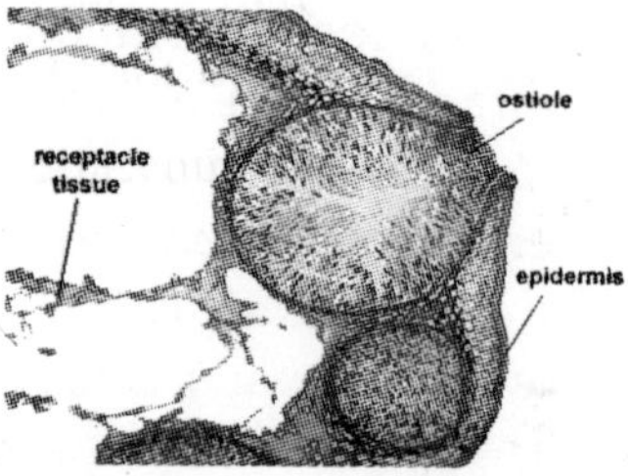

- **ostiolum**
  see **ostiole**.

- **outbreeding**
  the interbreeding of distantly related or unrelated individuals.

- **outcross**
  1. a cross between relatively unrelated individuals.
  2. the offspring of an outcross.

- **outcrossing**
  the contamination of a seed production field resulting from pollen of an undesirable plant strain pollinating the stigma of the parent seed plant.

- **ovary**
  the swollen bottom part of the pistil that contains the ovules or immature seeds.

- **ovate**
  egg-shaped, with the broader end at the base.

- **oven-dry soil**
  soil that has been dried at 105° C until it reaches constant mass.

- **over potting**
  putting a plant into a pot that is too big, often stops a plant flowering or becoming well established.

- **overhead watering**
  watering from above, usually with a sprinkler or a

hose with a sprinkler attachment.

- **overseeding**
  planting on top of an existing garden or lawn.

- **overstocked**
  1. refers to a stand that contains more trees or other materials than the site can support, causing tall, weak growth and the tendency to blow down.
  2. refers to a range which has more wildlife or domes-

tic stock than the area can support.

- **overtopping**
  when one branch of a dichotomy grows more than the other.

- **overwinter**
  1. to survive the winter.
  2. to keep alive through winter, e.g., bringing a soft plant indoors, wrapping roses or mulching pansies

- **ovule**
  the body which, after fertilisation, becomes the seed.

- **ovuliferous**
  bearing ovules.

- **oxic**
  containing oxygen, aerobic. Usually used in reference to a microbial habitat.

- **oxidation**
  process by which a compound gives up electrons, acting as an electron donor, and becomes oxidised.

- **oxidation-reduction (redox) reaction**
  coupled pair of reactions, in which one compound becomes oxidised, while another becomes reduced and takes up the electrons released in the oxidation reaction.

- **oxidative phosphorylation**
  synthesis of ATP involving a membrane-associated electron-transport chain and the creation of a proton-motive force. Also called electron-transport chain phosphorylation.

OXIDATIVE PHOSPHORYLATION

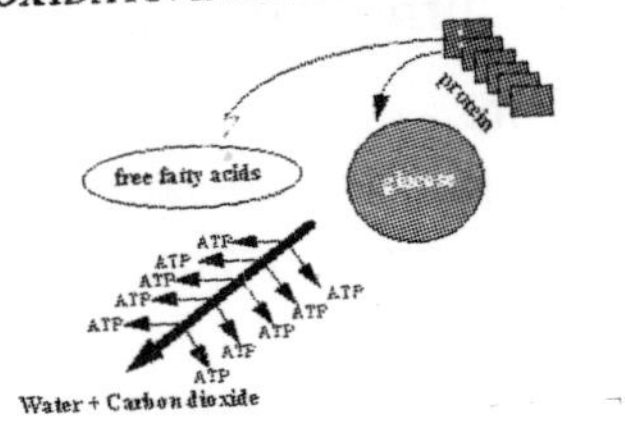

- **oxygen, 0^2,**
  a colourless, tasteless, odourless gas that is $^1/_5$ of the volume of the atmosphere.

- **oxygenic photosynthesis**
  use of light energy to synthesise ATP and NADPH by non-cyclic photophosphorylation with the production of oxygen from water.

- **pachycarpous**
  describes a seed having a thick outer layer or pericarp, e.g., the skin on a cherry. The skin (epicarp, ectocarp, or exocarp), flesh

(mesocarp or sarcocarp), and stony outer layer of the seed (endocarp) form a pachycarpous covering of the seed proper.

- **pachycaul**
with thick or massive primary construction.

- **pachydermous**
thick-skinned, applied to the walls of capsules or to cells when firm and resisting.

- **pack ice**
a large area of ice driven closely together.

- **paddock**
a grazing area that is a subdivision of a grazing management unit, and is enclosed and separated from other areas by a fence or barrier.

- **paecilomyces**
commonly found in soil and dust, less frequently in air. Linked to wood-trimmers disease and humidifier associated illnesses. They are reported to be allergenic. Some members of this genus are reported to cause pneumonia. It may produce arsine gas if growing on arsenic substrate.

- **page**
polyacrylamide gel electrophoresis.

- **pagoda**
a tower of far eastern design usually with roofs curving upward at the division of each of several stories and erected as a temple, a similarly styled garden structure.

- **paired**
refers to flowers or leaflets in opposite pairs, but bi-, tri-and multi-jugate further describe leaflets with two, three, or many such pairs or parts.

- **palate**
a rounded projection of the lower lip of a personate corolla, closing the throat.

- **palea**
the upper bract which, with the lemma, encloses the flower in grasses.

- **paleaceous**
  clothed with chaff.

- **paleobotany**
  the study of plant fossils.

- **paleophytological**
  relates to the study of fossil plants.

- **palisade**
  refers to the layer of columnar photosynthetic cells in the leaf.

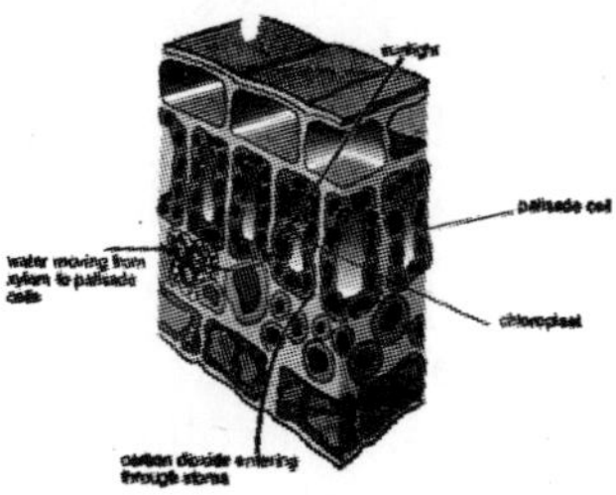

- **palm**
  any of various plants of the family Palmae, mostly tropical or subtropical monocotyledonous trees, shrubs, or vines, usually having a simple stem and a terminal crown of large pinnate or fan-shaped leaves.

- **palmate**
  describes a leaf that is radically lobed or divided.

A "palmate" leaf shape is like a hand with your fingers spread out. The main leaf veins start at one point & spread out into the leaf.

- **palmate leaf**
  five or more lobes arising from one point, hand like.

- **palmately compound**
  having veins or leaflets arranged like the fingers on a hand.

- **palmate-veined**
  with the principal veins arising from the end of the leafstalk and radiating toward the edge of the leaf.

- **palmatifid**
  describes a leaf which is deeply, but not completely, divided into several lobes.

- **palmatisect**
  intermediate between palmate and palmatifid.

- **palustrine**
  refers to areas that are not part of a major lake, but are filled with partially decomposed plant materials to considerable depth in wet-

lands such as bogs, swamps, and marshes.

- **palustrine wetlands**
those not a part of a main lake, therefore not subject to intensive wave or current effects.

- **palynology**
the scientific study of pollen and spores.

- **pan**
a layer of different soil below the surface, often impervious, which would be called hard pan.

- **panacea**
a remedy for all ills or difficulties, cure-all.

- **pandurate**
fiddle-shaped, with a rounded base, narrow waist, rounded upper part and long neck.

- **panicle**
a loose, irregularly compounded inflorescence with pedicellate flowers.

- **paniculate**
having the form of a panicle.

- **pannose**
like felt in texture, densely covered with hairs.

- **pantoporate**
describes a pollen grain having rounded apertures all over the surface.

- **pantropical**
growing throughout the warmer regions of the world. Used especially of the ranges of widespread weeds.

- **papilionaceous**
describes a corolla having standard wings, and keel, as in the peculiar corolla of many Leguminosae.

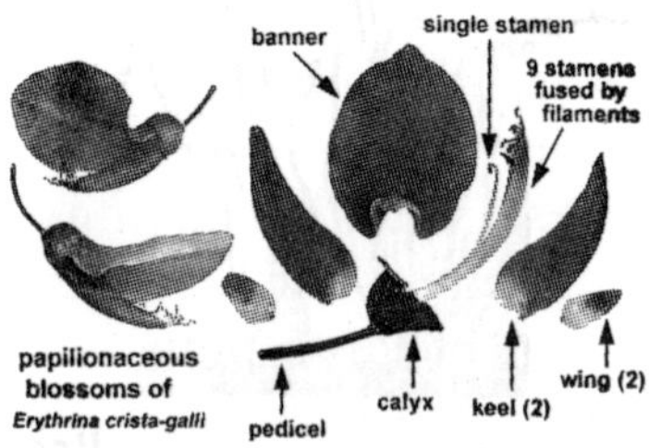

- **papilla**
1. a minute, nipple-shaped projection.
2. small bumps appearing on either the upper cortex or lower surface of various lichens.

• **papillaer**
minute rounded or acute protuberances.

• **papillate**
bearing papillae.

• **pappus**
the modified calyx-limb in Composite, forming a crown of various character at the summit of the achene.

• **papule**
a relatively large pustule, papilla, or pimple.

• **papulospora**
this fungi is found in soil, textiles, decaying plants, manure, and paper.

• **papyraceous**
paper-like.

• **paraclade**
an inflorescence on a lateral axis which repeats the symmetry of the primary axis.

• **paradichlorobenzene**
a white crystalline compound used in moth balls and moth crystals and often placed with herbarium specimens to prevent destruction by moth larvae. Large quantities in a closed container may be used to kill insects needed as part of sample.

• **parallel-ribbed**
with the veins running more or less parallel toward the tip of the leaf.

• **paraphyllia**
minute leaf-like or much-branched organs among the leaves.

• **paraphysis**
sterile, elongated cell that may occur in the hymenium, intermixed with asci or basidia, elongating apically and having a free apex.

• **parasexual cycle**
a sequence involving heterokaryon formation, diploidisation, and haploidisation, often resulting in the formation of recombinant nuclei. Unlike the sexual cycle, the parasexual cycle can occur at

any point or continuously throughout the life cycle

- **parasite**
  an organism living in or on another living organism. (host from which it extracts nutrients.)

- **parasitisation**
  the process whereby one organism lives on or with another as a parasite.

- **parataxonomists**
  field-trained biodiversity inventory specialists collecting in and recruited from local areas.

- **paratype**
  a specimen other than the holotype and its isotypes that the author cited at the time of publication of the original description.

- **parenchyma**
  a tissue composed of living, thin-walled cells that can continue to divide even when mature, parenchyma cells usually leave intercellular spaces between them.

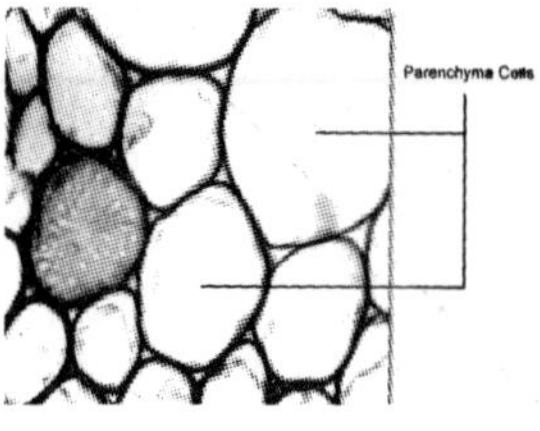

- **parenchymatous**
  cells with broad ends abutting on each other, not dovetailing into each other.

- **parietal**
  borne on or pertaining to the wall or inner surface of a capsule.

- **paripinnate**
  having an even number of pinnae.

- **parkland**
  land in which clumps of trees are scattered throughout a grassland.

- **parks**
  areas set aside by government and kept in their natural states for the purpose of conserving unique areas for preserving game, walking, riding, or recreation.

- **paroicous**
  having its male and female organs in the same cluster, but not mixed, the antheridia being in the axils of the perichaetial bracts below the archegonia.

- **parted**
  cleft nearly but not quite to the base.

- **parterre**
symmetry at its best. These are great ornamental garden beds that have been geometrically designed and separated by walkways. The gardens are most appreciated by an above view.

- **parthenocarpic**
refers to the fruiting of plants which have not been pollinated or otherwise fertilised

- **parthenogenesis**
process of reproduction by the development of an unfertilised egg.

- **parthenogenetic**
seed which develops without fertilisation, but by stimulus only.

- **partial**
of secondary rank.

- **particle density**
density of the soil particles, the dry mass of the particles being divided by the solid (not bulk) volume of the particles, in contrast with bulk density.

- **particle size**
effective diameter of a particle measured by sedimentation, sieving or micrometric methods.

- **partite**
parted.

- **partitioned**
the pith divided crosswise by woody plates, usually near the leaf scars.

- **passalong plants**
plants that are shared between friends and not easily found in catalogs. Usually some plant from an older garden.

- **pasteurization**
process using mild heat to reduce microbial numbers in heat-sensitive materials.

- **pasture**
a type of grazing management unit enclosed and separated from other areas by fencing or other barriers and devoted to the production of forage for harvest primarily by grazing.

- **pasture carrying capacity**
  see **carrying capacity**.

- **pasture, supplemental**
  a crop used to provide grazing for supplemental use, usually during periods of low pasture production.

- **pasture, temporary**
  a field of crop or forage plants grazed for only a short period, usually not more than one crop season.

- **patellate**
  round and thick, with one side concave and the other following the same curve by being convex, like a thick contact lens.

- **patent**
  spreading, as do the branches of a tree, particularly at an angle of 26 to 45 degrees.

- **path coefficient**
  a measure of the importance of a given path of influence from cause to effect, it is the ratio of the standard deviation of the effect (with all causes kept constant except the one in question, the variability of which is kept unchanged) to the total standard deviation. This is the standard partial regression coefficient in a prediction equation in which all independent variables are causes of the effect and all relevant variables are included.

- **pathogen**
  organism able to inflict damage on a host it infects.

- **pathogenesis**
  that portion of the life cycle of a pathogen during which it becomes, and continues to be, associated with its suscept.

- **pathogenic**
  having the characteristics of a pathogen.

- **pathogen-suppressive soil**
  soil where a pathogen does not establish or persist, a pathogen establishes but causes little or no damage, or a pathogen causes disease for a while, but the disease becomes less impor-

tant even though the pathogen persists in soil.

- **pathology**
  1. the study of disease.
  2. the abnormal condition that constitutes disease.

- **pathotype**
  an infra-subspecific classification of a pathogen distinguished from others of the species by its pathogenicity on a specific host(s).

- **pathovar**
  in bacteria: An infra-subspecific group that can infect only plants within a certain genus or species.

- **patina**
  the changed colour or texture of a surface due to weathering or aging.

- **patulous**
  more widely spreading than patent.

- **PCR**
  see **Polymerase Chain Reaction**.

- **pea gravel**
  gravel about the size of a pea. Used a lot in driveways and walkways.

- **pea-like**
  refers to a flower with a top banner or standard, similar to legume blossoms, papilionaceous.

- **peat**
  unconsolidated soil material consisting largely of nondecomposed, or only slightly decomposed, organic matter accumulated under conditions of excessive moisture.

- **peat bog**
  an area with a wet, spongy, acidic substrate, consisting mainly of accumulated layers of sphagnum moss and upon which other plants, and even trees, can grow.

- **peat moss**
  partially decomposed sphagnum moss, often added to soil to increase moisture retention.

Sphagnum Peat Moss

- **peat pot**
compressed peat into a pot that can be used for starting seeds. When planting times comes this entire pot can be put in the ground and the roots will grow through the pot as it decomposes.

- **peat substitute**
because of environmental concerns over the use of peat, a substitute is now available made from renewable materials like coir.

- **peavey**
a tool similar to a cant hook, but with the tip outfitted with an iron spike rather than a ring.

- **pebble tray**
grouping potted plants within a shallow, pebble filled tray in order to maintain humidity in an environment with central heating. Water is poured into the pebbles and evaporates up and around the plants.

- **pectin**
any of various water-soluble substances that bind adjacent cell walls in plant tissues and yield a gel which is the basis of fruit jellies.

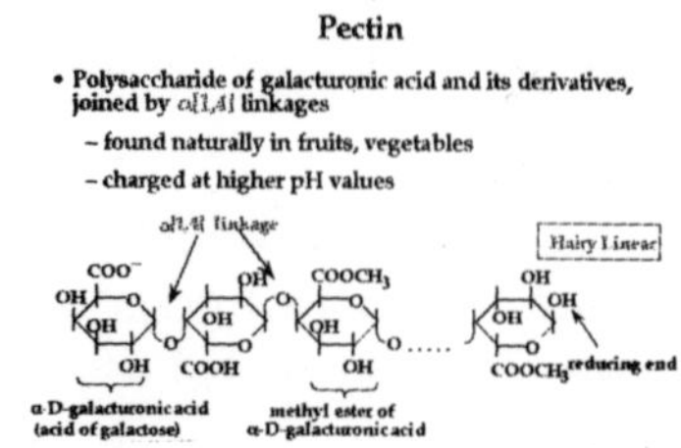

- **pectinate**
pinnatifid with narrow, closely set segments, comb-like.

- **pedate**
palmately divided or parted, with the lateral segments to cleft.

- **pedicel**
1. a tiny stalk, the support of a single flower.
2. the stalk of the sporangia.

- **pedicellate ovary**
the combined pedicel with pedicellate of the flower

- **pedicelled**
borne on a pedicel.

- **pedigree**
1. a system of plant breeding in which detailed records are kept of all plants in segregating generations following hybridisation.
2. a record, using standardised notation, showing the line descent for a variety.

- **peduncle**
the stalk of an flower head.

- **pegging**
securing the ends of canes to the ground so that the plant grows horizontally.

- **pegging down**
a very effective way to increase flower production on Bonbon & Hybrid Perpetual roses that send up long shoots with oily flowers on the end. It is a time consuming method but well worth the display. This is generally done in the fall. Instead of pruning, spread the runners and fasten to the soil with wire loops. Basically you are training the plant horizontally.

- **pellet**
the material concentrated at the bottom of a centrifuge tube after centrifugation

- **pelleted seeds**
seeds that have been coated with an inert material just

to make the handling of the seed easier.

- **pelletised**
the coating and forming into pellets of very small seed so they are easier to handle.

- **pellicle**
relatively rigid layer of proteinaceous elements just beneath the cell membrane in many protozoa and algae.

- **pellucid**
clear, transparent

- **peloderan**
in nematodes: Caudal alae that meet posterior to the tail tip.

- **peloria**
unusual regularity or symmetry in the form of a

flower that is normally irregular.

- **peltate**
  1. describes a leaf attached to the petiole from near the centre of the lower surface, and not at the margin.
  2. shield-shaped

- **pendant**
  hanging downward.

- **pendulous**
  more or less hanging or declined. Pendulous ovule: one that hangs from the side of the cell.

- **penetration**
  the point at which a pathogen enters a host.

- **penetration peg**
  in some plant parasitic fungi: the peg-like hypha emerging from an appressorium that penetrates the epidermal cell wall

- **penicillium**
  a wide number of organisms have placed in this genera. Identification to species is difficult. Often found in aerosol samples. Commonly found in soil, food, cellulose and grains. It is also found in paint and compost piles. It may cause hypersensitivity pneumonitis, allergic alveolitis in susceptible individuals. It is reported to be allergenic (skin). It is commonly found in carpet, wallpaper, and in interior fibreglass duct insulation. Some species can produce mycotoxins. Acute symptoms include edema and bronchiospasms, chronic cases may develop pulmonary emphysema.

- **penniribbed**
  having conspicuous lateral veins which are divergent from the midrib and approximately parallel to one another.

- **pentamerous**
  describes a flower which has five parts in each floral whorl.

- **pentangular**
  with five angles, as the cross-section of a stem.

- **pentaploid**
  having five sets of chromosomes.

- **pepo**
  a fruit, such as that of a melon or squash, having a firm rind, fleshy pulp, many flattened seeds, and a single locule.

- **peptidoglycan**
  rigid layer of cell walls of bacteria, a thin sheet composed of N-acetylglucosamine, N-acetylmuramic acid, and a few amino acids. Also called murein.

- **perched water table**
  the water table of a smaller body of groundwater body situated above the general groundwater table.

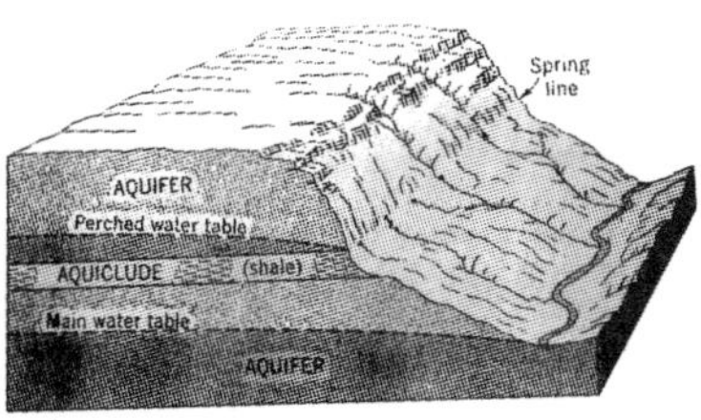

- **percurrent**
  describes a Costa that reaches to the apex of the leaf, but not beyond.

- **perennate**
  to live for more than one growing season, but usually with reduced growth or dormancy between growing seasons.

- **perennial**
  a plant which will live for three years or more under normal conditions.

- **perennial ryegrass**
  a cool-season turf grass with seeds that germinate quickly. The cultural requirements are similar to those of Kentucky bluegrass; however, it is not quite as hardy or disease resistant as bluegrasses.

- **perfect**
  describes a flower having both pistil and stamens.

- **perfect state**
the state of a fungus characterised by sexual spores.

- **perfoliate**
with the leaf surrounding the stem.

- **perforate**
pierced with holes.

- **pergola**
an open structure usually consisting of parallel columns supporting an open roof of cross rafters on which climbing plants are .rained to grow, an arbour.

- **perianth**
the floral envelope, consisting of the calyx and corolla (when present), whatever their form.

- **peribacteroid membrane**
plant-derived membrane surrounding one to several rhizobia within host cells of legume nodules.

- **pericarp**
the wall of the matured ovary.

Brown pericarp and seed coat

- **perichaetium**
a whorl of bracts at the base of reproductive organs. In mosses, those surrounding the archegonia and base of seta.

- **pericycle**
the tissue of the stele lying just inside the endodermis.

- **periderm**
an outer, cortical protective layer of many roots and stems that typically consists ot phellem, phellogen, and phelloderm.

- **peridium**
a wall or membrane of sterile cells around a fruiting body (e.g., around a sporangium or delimiting an aecium (pl. peridia)).

- **perigone**
  the perianth, especially when made up of tepals, or in reference to anything surrounding the reproductive structure.

- **perigonium**
  1. the perianth.
  2. in mosses, those bracts surrounding the antheridia.

- **perigynium**
  the inflated sac which encloses the ovary in Carex.

- **perigynous**
  adnate to the perianth, and therefore around the ovary and not at its base.

- **perilite**
  a mineral expanded by heating to form very lightweight, porous white granules useful in container soil mixes to enhance moisture and air retention.

- **perineal pattern**
  fingerprint-like pattern formed by cuticular striae surrounding the vulva and anus of the mature Meloidogyne female.

- **period of occupation**
  the length of time that a given land area is occupied, whether by one animal group, or by two or more animal groups in succession. Usage: Period of occupation may involve several different groups moving through in sequence, (as in first-last grazing or in migrations). It differs from grazing period in that grazing may or may not be involved (e.g., feeding hay on pasture in winter when ice prevents grazing).

- **periodicity**
  repetition of events at fairly regular intervals.

- **peripheral**
  on or near the margin.

- **periphysis**
  short, hair-like filaments that line the canal of the ostiole in some Pyrenomycetes. (pl. periphyses)

- **periphyton**
  the assemblage of organisms submerged in water, attached to surfaces above the bottom of the body of water.

• **periplasmic space**
area between the cell membrane and the cell wall in Gram-negative bacteria, containing certain enzymes involved in nutrition.

• **perisperm**
the nutritive tissue in an angiosperm seed that surrounds the embryo, formed from the nucellus.

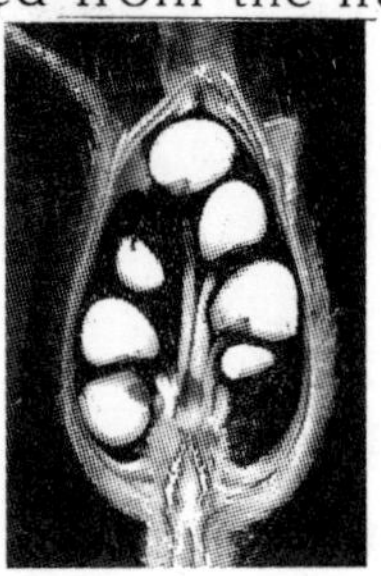

• **peristome**
the fringe surrounding the mouth of the capsule upon removing the lid.

• **perithecium**
flask-shaped ascocarp open at the tip, containing asci of fungi in the phylum Ascomycota.

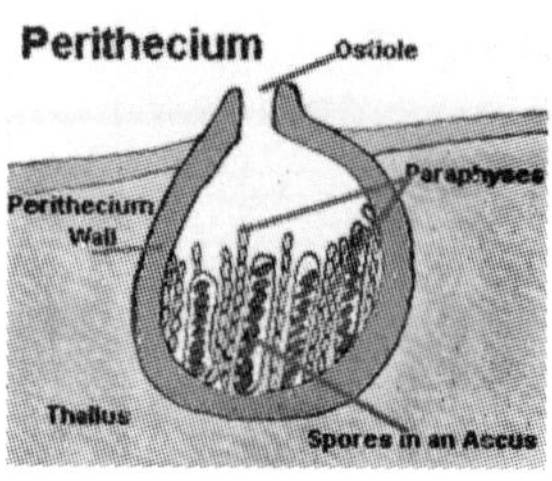

• **peritrichous**
of bacterial flagella: distributed more or less uniformly over the cell surface.

• **perlite**
a lightweight aggregate made from a volcanic glass that has been expanded by heat to form white, very lightweight kernels useful in opening cavities in soil to allow water and air to reach the roots, often used in potting soil.

• **perloline**
a plant alkaloid found to interfere with cellulose digestion by rumen micro-organisms, commonly associated with tall fescue.

• **permaculture**
the maintaining of a permanent horticulture or agriculture by relying on renewable resources and compatibility with the local ecosystem.

• **permafrost**
frozen ground in arctic and subarctic areas which does not thaw in summer.

• **permanent wilting point**
greatest water content of a soil at which indicator

plants, growing in that soil wilt and fail to recover when placed in a humid chamber. Often estimated by the water content at -1.5-MPA soil metric potential.

- **peronosporales**
specialised forms of the Oomycetes, including aquatic and terrestrial species. Many species in this order are plant pathogens (damping-off) fungi, downy mildews, and white rusts. Unlike the true fungi, they lack chitin in their cell walls.

- **persistent**
1. retaining its place, shape, or structure, remaining attached after the growth period, not deciduous, designating fruits, flowers, leaves, etc., that remain on a plant after frost.
2. of a disease: difficult to control.

- **personate**
describes a corolla which is bilabiate, and the throat closed by a prominent palate.

- **perula**
1. one of the scales of a leaf bud.
2. a pouch like portion of the perianth in certain orchids

- **perulate**
describes leaf buds which are covered with scales.

- **pesticide**
a chemical that kills pests.

- **petal**
a whorl of structures that surround the inner reproductive organs of a flower. Together they are called the corolla. They often attract insects by colour or nectar, facilitating pollination.

- **petaliferous**
bearing petals

- **petaline**
of or resembling petals.

- **petaloid**
1. coloured and resembling a petal.
2. additonal floral parts on ray florets having the form and appearance of smaller

petals, e.g., in the Collarette dahlias.

- **petiolate**
  having a petiole.

- **petiole**
  leaf stock, slender stalk by which a leaf is attached to the stem.

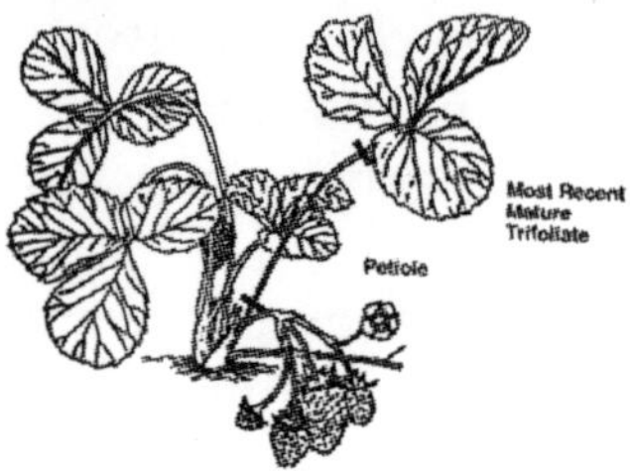

- **pH**
  negative logarithm of the hydrogen ion activity. The degree of acidity (or alkalinity) of a soil as determined by means of a glass or other suitable electrode or indicator at a specified moisture content or soil-water ratio, and expressed in terms of the pH scale.

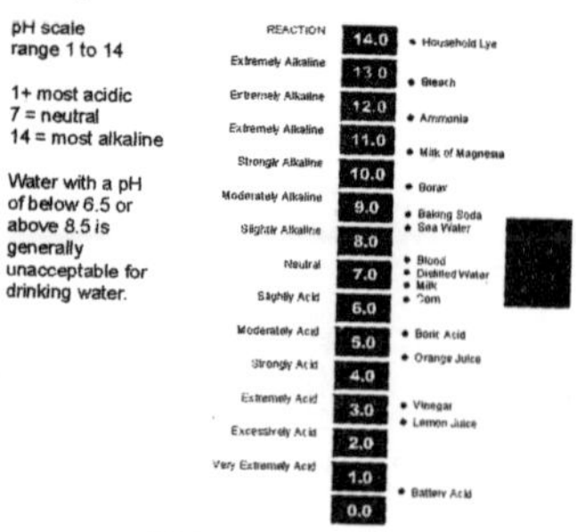

- **phaenogamous**
  having flowers with stamens and pistils and producing seeds.

- **phage**
  a general term used for viruses isolated from prokaryotes including bacteria, blue-green algae (cyanobacteria) and mollicutes (phytoplasma and spiroplasma). The viruses from these different host groups are termed bacteriophages, cyanophages and mycoplasmaphages, respectively.

- **phagotrophic**
  form of feeding where animals, such as protozoan, engulf particulate nutrients, such as bacterial cells or detritus.

- **phalaenopsoid**
  growing like a Phalaenopsis.

- **phanerogam**
  a general name for flowering plants.

- **phaneropore**
  superficial stoma.

- **phasmid**
  in nematodes: a pore-like structure located in the lat-

eral field of the posterior region of nematodes belonging to the class Secernentea. Function is believed to be sensory. Sometimes called precaudal glands.

- **phellem**
  a layer of usually suberised cells produced outwardly by a phellogen.
- **phelloderm**
  a layer of parenchyma produced inwardly by a phellogen.
- **phellogen**
  a secondary meristem that produces phellem and phelloderm in the periderm of a trunk or stem.
- **phenetic classification**
  the grouping of taxon by apparent similarities rather than evolutionary genetics.
- **phenology**
  the science of the relations between climate and periodic biological phenomena, e.g., the fruiting of plants or the colour change of leaves
- **phenotype**
  the morphological, physiological, behavioural, and other outwardly recognisable adaptations of an organism that develop through the interaction of genes and environment.
- **phenotypic**
  refers to a plant's adaptation to surrounding conditions, which are neither stable nor capable of being inherited (genotypic). Such visible changes occur especially where plants are grown in a wide variety of conditions, but will not carry over to different conditions, e.g., red leaves may occur in hot dry areas, but turn green when grown in normal conditions.
- **pheromones**
  chemical substances produced by animals that attract and stimulate sexual partners of the same species.
- **phloem**
  the part of the vascular system that moves food through the plant.

COMPONENTS OF PHLOEM

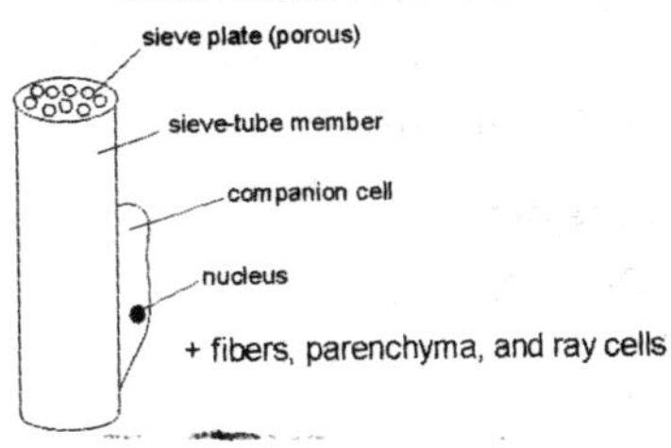

- **phoma**
  a common indoor air allergen. It is similar to the early stages of growth of Chaetomium sp. The species are isolated from soil and associated plants (particularly potatoes). Produces pink and purple spots on painted walls. It may have antigens, which cross-react with those of Alternaria. It will grow on butter, paint, cement and rubber. It may cause phaeohyphomycosis-a systematic or subcutaneous disease.

- **phosphobacterium**
  bacterium that is especially good at solubilising the insoluble inorganic phosphate in soil.

- **phosphodiester bond**
  type of covalent bond linking nucleotides together in a polynucleotide.

- **phospholipid**
  lipids containing a substituted phosphate group and two fatty acid chains on a glycerol backbone.

- **phosphorus cycle**
  sequence of transformations undergone by phosphorus where it is transformed between soluble and insoluble, and organic and inorganic forms.

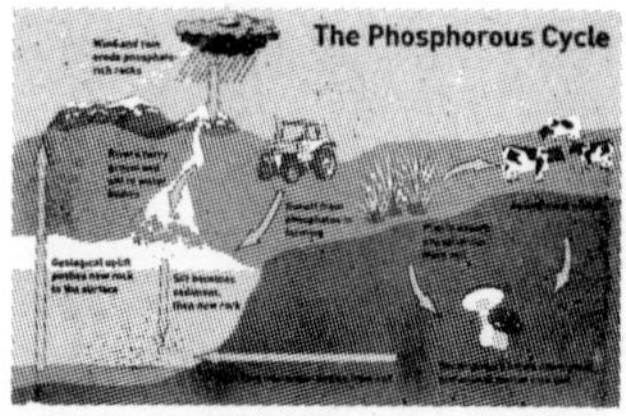

- **photic zone**
  the upper layers of bodies of water into which sunlight penetrates sufficiently to influence the growth of plants and animals.

- **photoautotroph**
  organism able to use light as its sole source of energy and carbon dioxide as sole carbon source.

- **photoheterotroph**
  organism able to use light as a source of energy and organic materials as carbon source.

- **photomania**
  the response of an organism of seeking or growing toward light.

- **photomorphogenesis**
  a special case of morphogenesis in which light, by virtue of absorption of particular wavelengths, promotes de-

velopment of tissues or organs. Photomorphogenetic pigment systems commonly recognised are: (a) phytochrome, (b) a blue-absorbing high-energy system perhaps related to phytochrome, and (c) a riboflavin-carotenoid system active in phototropism.

- **photoperiod**
  the duration of an organism's daily exposure to light.

- **photoperiodism**
  the response of plants to the length of a day and night (i.e. poinsettias, cactus, night blooming flowers).

- **photophobia**
  the dislike of light, as displayed by bugs that hide under rocks or bats which sleep all day and fly at night.

- **photophosphorylation**
  the synthesis of ATP from ADP and phosphate that occurs in a plant using radiant energy absorbed during photosynthesis.

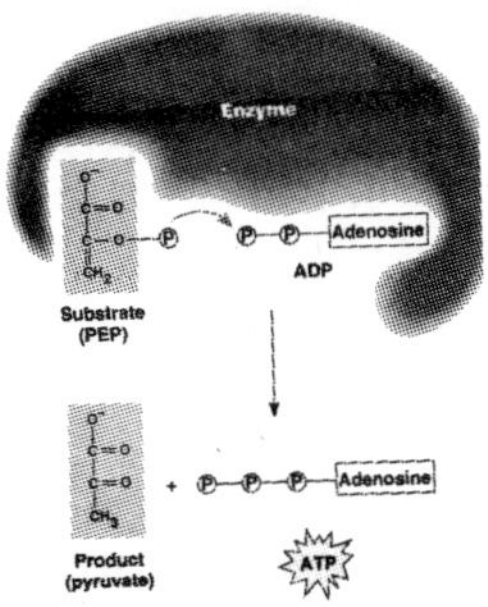

- **photosensitisation**
  a non-contiguous disease in animals resulting from the abnormal reaction of light-coloured skin to sunlight after a photodynamic agent has been absorbed through the system. Grazing certain kinds of vegetation or ingesting certain moulds under specific conditions causes photosensitisation.

- **photosynthesis**
  the production of sugar from carbon dioxide and water in the presence of chlorophyll, activated by light energy and releasing oxygen.

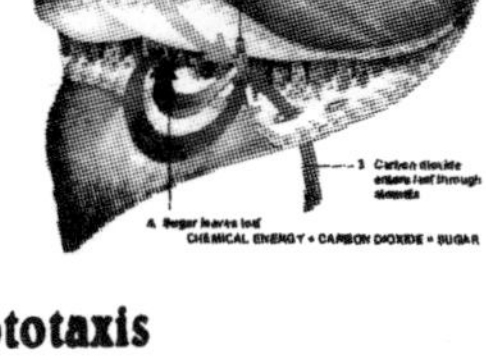

- **phototaxis**
  the movement of a body toward or away from a light source.

- **phototroph**
organism that uses light as the energy source to drive the electron flow from the electron donors, such as water, hydrogen, or sulphide.

- **phototropism**
the bending of a plant toward the direction of more intense light.

- **phragmobasidium**
a basidium that is divided into more than one cell by transverse or longitudinal septa.

- **phragmospore**
a spore having two to many transverse septa.

- **phreatophyte**
a plant that can endure inundations of salt, such as cattails, Typha, which can live in estuaries, sieving out the saline molecules of seawater at a cellular level.

- **phycobilin**
water-soluble pigment that occurs in cyanobacteria and functions as the light-harvesting pigments for Photosystem II.

- **phyllary**
one of the bracts under the flower head of a plant, especially in Compositae.

- **phylloclade**
a flattened, photosynthetic stem that performs the functions of a leaf, as occurs on some cacti.

- **phyllode**
a flat expanded petiole that replaces the blade of a foliage leaf and fulfills the same functions.

- **phyllodium**
a somewhat dilated petiole having the form of and serving as a leaf blade.

- **phyllody**
a change of floral petals leaves to foliage leaves

- **phylloplane**
the surface(s) of a leaf.

- **phyllopodium**
an outgrowth of the rhizome to which the frond is joined in some ferns.

- **phyllotaxy**
the manner of leaf arrangement on a stem.

- **phylogenetic classification**
the grouping of taxon by genealogical descent, evolution.

- **phylogeny**
ordering of species into higher taxon and the con-

struction of evolutionary trees based on evolutionary (genetic) relationships.

- **phylum**
  a major taxonomic grouping in the animal kingdom, ranking just below kingdom and above class. In the plant kingdom, the division usually replaces it.

- **physiographic climax**
  a pinnacle habitat controlled by the topography of the area, e.g., a forest growing on a north slope and a grassland on the south slope of the same ridge.

- **physiologic specialisation**
  the existence of a number of races or forms of one species of pathogen based on their pathogenicity to different cultivars of a host.

- **physiological race**
  a subdivision of a species of pathogen, particularly fungi, distinguished from other members of the species by specialisation for pathogenicity to different cultivars of a host.

- **phytoalexin**
  a low molecular weight, ant microbial compound synthesised by and accumulating in higher plants exposed to certain micro organisms (pathogenic and non-pathogenic).

- **phytochemistry**
  the chemistry of plants, plant processes, and plant products.

- **phytochrome clocks**
  the colouring processes of plants that change the hues of ripening fruits and cause leaves to change colours with the shortened days of autumn.

- **phytogenic**
  having a plant origin, e.g., coal.

- **phytography**
  the science of plant description.

- **phytology**
  the study of plants.

- **phytopathogenic**
  of micro organisms: capable of initiating disease in plants.

- **phytoplankton**
  small, often microscopic, aquatic plants.

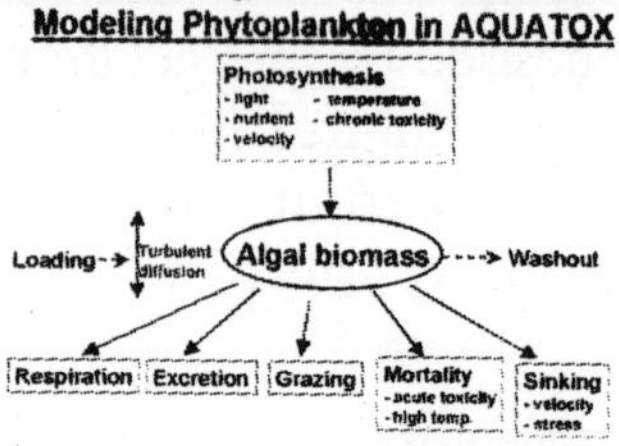

- **phytoplasma**
  a prokaryotic, plant parasitic micro-organism resembling a mycoplasma but not yet isolable in pure culture or characterised taxonomically.

- **phytotoxic**
  toxic to plants.

- **phytotoxin**
  1. a toxin produced by a micro-organism and active against a plant or against plant cells/tissues.
  2. a toxin produced by a plant.

- **pick**
  a mattock type tool, which has one or both ends pointed.

- **pick pruning**
  the selection of individual branches to be cut, the opposite of shearing.

- **picotee**
  term applied to a narrow band of colour on a pale ground at the edge of a petal.

- **pier**
  a mass of masonry used as a support, breakwater, etc.

- **pier hole**
  a landscape term for a deep narrow hole.

- **pileate**
  possessing a cap or pileus.

- **pileus**
  the expanded cap like portion of some basidiocarps or ascocarps that supports the hymenium (pl. pilei)

- **pilose**
  hairy, especially with soft hairs.

- **pilosulous**
  minutely pilose.

- **pilus**
  fimbria-like structure that is present on fertile cells and is involved in DNA transfer during conjugation. Sometimes called sex pilus. (plural pili)

- **pinch**
  breaking off the terminal growing point of a plant to encourage axillary buds to grow.

- **pinch back**
  to remove the growing tips on main stems, usually using the fingernails of thumb and forefinger, in order to induce branching and thereby thicken and strengthen the plant.

- **pinching out**
  the removal of the growing point of a stem to induce bushiness or to encourage flowering. Also known as stopping.

- **pinna**
  one of the primary divisions of a pinnate or compoundly pinnate frond or leaf.

- **pinnate**
  1. consisting of several leaflets arranged on each side of a common petiole or rachis on a compound leaf or frond.
  2. the feather vein pattern of simple leaves.

- **pinnate leaf**
  a series of leaflets arranged on either side of a central stalk.

- **pinnately compound**
  with leaflets arranged in two rows along an axis.

- **pinnatifid**
  divided in a pinnate manner, but with leaflets not entirely separate.

- **pinnatisect**
  deeply cut, all the way to the axis.

- **pinnule**
  a secondary pinna, one of the pinnately disposed divisions of a pinna.

- **pip**
  1. the small seed of a fruit, like that of an apple or an orange.
  2. an individual rootstock of lily of the valley or a similar plant.

- **pistil**
  the female or seed-producing organ of a flower, consisting usually of the ovary, style, and, in orchid. The pis-

til becomes part of the column and pedicellate ovary

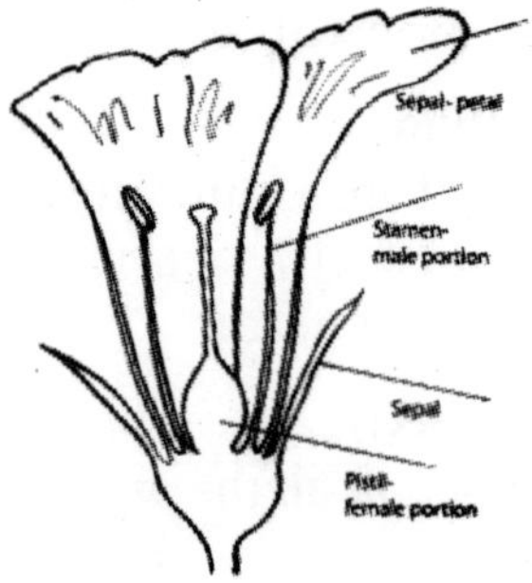

Diagramatic flower structures - cut in half

- **pistillode**
  a sterile pistil, often rudimentary.

- **pistilloidy**
  the conversion of any organ of a flower into carpels, often stamens into pistils or pistil like structures.

- **pitch**
  lumps of resin found on the bark of trees.

- **pitcher**
  any tubular or cup-like vessel, most often a modified leaf, generally used to trap insects which provide nitrogen to a plant growing in an acid environment.

- **pitcher-shaped**
  campanulate, but with a distinct narrowing near the open end.

- **pith**
  the spongy or hollow centre of twig or some stems.

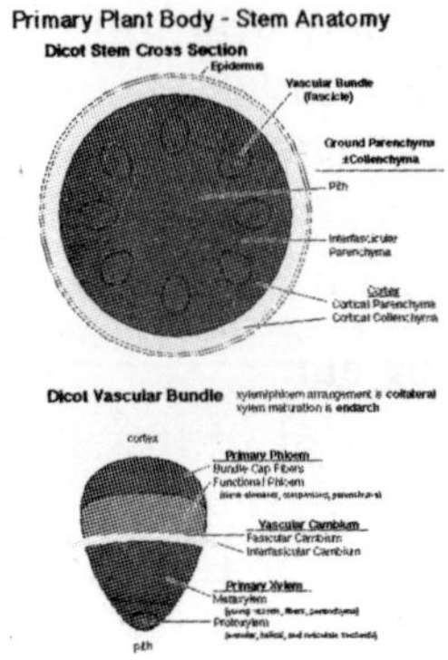

- **pithomyces**
  grow on dead grass in pastures. Causes facial eczema in ruminants.

- **pitted**
  marked with small depressions or pits.

- **placenta**
  any part of the interior of the ovary which bears ovules.

- **placentation**
  the arrangement of placentas within an ovary.

- **plankton**
  the floating or weakly swimming plants and animals occurring at any depth in bodies of water, often microscopic in size.

- **plant breeder**
  person or organisation actively engaged in the breeding and maintenance of varieties of plants.

- **PGPR**
  Plant Growth-Promoting Rhizobacteria. Broad group of soil bacteria that exert beneficial effects on plant growth usually as root colonisers.

- **plant labels**
  1. plastic, wood, or metal stakes for gardens to indicate what seeds are planted where until they appear, and for varieties for evaluation. 2. paper forms to include in drying plant samples, with formal printed forms as permanent labels on herbarium specimens. The minimum information required are the name of the collector, the location collected, the date collected, and hopefully, the correct identification of the specimen.

- **plant lice**
  this is a reference to aphids found in British publications.

- **plant patent numbers**
  this generally is a catalogue referral phrase. For the general gardener it may not be important but new plants, like inventions, can also be patented. This is a protection for the owner who created the 'new' plant.

- **plant press**
  a structure typically made of two ventilated frames 12 X 18 inches (30 X 45 cm.), within which plants may be arranged between sheets of driers and ventilators, tightly strapped together with press straps, in preparation of being added to a collection or herbarium.

- **plant window**
  double window with plants grown in the space between.

- **plantation**
  a large farm of cultivated trees or plants, often of a single crop.

- **planting density**
  the rate at which seed or vegetative propagules are placed in a field or nursery planting.

- **plantlet**
  a small plant off the original plant. A good example is the piggyback plant. These will

easily root. Used in propagation.

- **plaque**
  localised area of lysis or cell inhibition caused by virus infection on a lawn of cells.

- **plasma membrane**
  see **cytoplasmic membrane**.

- **plasmalemma**
  the cytoplasmic membrane found on the outside of the protoplast adjacent to the cell wall.

- **plasmid**
  in many types of prokaryotic and eukaryotic cell: a linear or covalently closed circular molecule of DNA, (distinct from chromosomal DNA), mtDNA, ctDNA, or kDNA and commonly dispensable to the cell, that can replicate autonomously (i.e., independently of other replicons).

- **plasmodesma**
  a fine protoplasmic thread connecting two protoplasts and passing through the wall separating the two protoplasts. (pl. plasmodesmata)

- **plasmodium**
  a multi nucleated, usually naked (i.e., bounded only by a plasma membrane) mass of protoplasm that is usually motile and variable in size and form. (pl. plasmodia).

- **plasmogamy**
  fusion of the contents of two cells, including cytoplasm and nuclei.

- **plasmolysis**
  the shrinking and separation of the cytoplasm from the cell wall due to exosmosis of water from the protoplast.

- **plasticity**
  flexibility, adaptability, capable of being repeatedly deformed without rupture.

- **plastid**
  any of various cytoplasmic organelles of photosynthetic cells that serve in many cases as centres of special metabolic activities, e.g., chloroplasts.

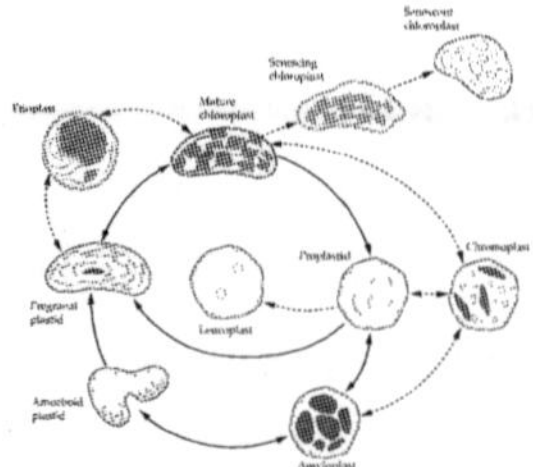

- **plate count**
number of colonies formed on a solid culture medium when uniformly innoculated with a known amount of soil, generally as a dilute soil suspension. The technique estimates the number of certain organisms present in the soil sample.

- **plates**
flattened, rhizine-like structures on the lower surface of some lichens.

- **pleaching**
a process whereby branches of woody plants are interwoven and plaited together to form an impassible hedge or very thick arbor. The only pruning done is to maintain a neat formal shape.

- **pleated**
describes a leaf creased along its length.

- **plectomycetes**
in general, a group of primitive or reduced ascomycetous forms that have an angiocarpous fructification without an ostiole, the entire interior of which is irregularly penetrated by ascogenous hyphae, with the result that the generally spherical asci, without accompanying paraphyses or other threads, lie scattered irregularly in a pseudoparenchymatous tissue composed of the ascogenous hyphae.

- **pleomorphic**
exhibiting pleomorphism.

- **pleomorphic**
able to assume different forms.

- **pleomorphism**
1. in fungi: having more than one independent form or spore stage in the life cycle.
2. in general: an inherent variability in size and shape (e.g., among the cells in a pure culture or clone of a given organism.

- **plerome**
the plant tissues inside the cortex.

- **plesionecrosis**
a symptom exhibited by tissues not yet dead but in the process of dying.

- **pleurocarpous**
having the sporophyte lateral on a short lateral special branch. Pleurocarpous

mosses can usually be recognised by the creeping habit.

- **plicae**
  folds of a plicate leaf.

- **plicate**
  folded into plaits, usually lengthwise.

- **ploidy**
  the number of (complete sets of) chromosomes in a cell.

- **plow**
  a device used to turn soil. On large farms several of these are mounted in a diagonal line, and referred to by the number of blades, e.g., a six-bottom plow. For compacted soil, chisels, four feet (1.3 m.) or longer, are used to open greater depth, and are called chisel plows.

- **plug**
  a small but well-rooted seedling raised in a cellular tray for covering large areas as in ground covers or lawns.

- **plugging**
  vegetative propagation of turf grasses by plugs or small pieces of sod. Used for establishing vegetatively propagated turf grasses or repairing damaged areas.

- **plume**
  a feathery inflorescence.

- **plumose**
  having fine hairs on each side, like the plume of a feather, as the pappus-bristles of some thistles.

- **plumule**
  the bud or growing point of the embryo.

- **plunge**
  to place an outdoor pot up to its rim in soil, peat, sand or ashes to protect the roots from extreme heat or cold.

- **plurilocular**
  describes a sporangium or gametangium composed of many cells, each producing one zoospore or gamete.

- **pluriseriate**
  many-ranked, as applied to leaves arranged in several rows along the stem.

- **pneumatophore**
  a specialised root in certain aquatic plants which performs respiratory functions.

- **pocket beach**
  the shore at a bay head where wave energy is lowest and settlings highest, often with fringe wetlands.

- **pocket garden**
  a small growing area planted with miniature and dwarf varieties.

- **pod**
  any dry, several-seeded and dehiscent fruit.

- **podetium**
  a hollow upright structure.

- **polar**
  at one end or pole of the cell (e.g., a flagellum, spore inclusion, germ tube, etc).

- **polar flagellation**
  condition of having flagella attached at one end or both ends of the cell.

- **polarity**
  the tendency for plants to develop from its poles, roots growing down, stems growing upward, making it essential to plant bulbs, etc, in the correct position.

- **pole pruner**
  a pruner on an extensible pole usually having a curved saw at its end, which can be used to prune the upper parts of a tree or tall shrub.

- **pollard**
  a tree cut back to the trunk to make a dense cluster of branches and foliage.

- **pollarding**
  a process where tree tops are cut back severely each year to the same spots on the branches. This forces the growth of large knobby stubs from which long tender shoots grow each year.

- **pollen**
  spores or grains borne by the anther, containing the male element. In orchids, it is usually not granular, as in most other plants.

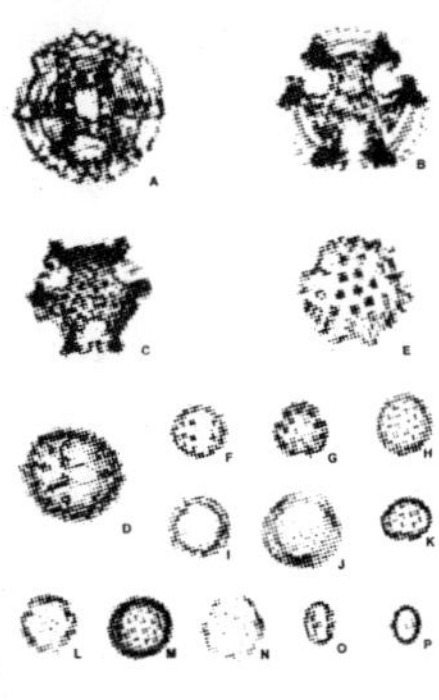

- **pollen culture**
  the in vitro culturing of pollen grains to obtain haploid callus or haploid plantlets directly from the pollen grains. Pollen culture differs from microspore cultures by the stage of development in gametogenesis cycle.

- **pollen grain**
  a microspore of a seed plant.

- **pollen sac**
  the microsporangium of a seed plant where pollen is produced, the upper portion of the stamen containing pollen grains, the anther.

- **pollen tube**
  the slender tube that is emitted by a pollen grain which penetrates and fertilises the ovule.

- **pollinarium**
  the apperatus of the orchid used to transport pollen from one flower to another. The pollination consists of The Pollinia (pollen packets), the Caudicle (a stalk-like appendage), and the Viscidium (a sticky gland).

- **pollinate**
  to transfer pollen from the anther of a stamen to the stigma of a pistil, resulting in fertilisation. This can occur either on a single plant (self-pollination) or between different plants.

- **pollination**
  the transfer of pollen from the anther to the stigma of a flower.

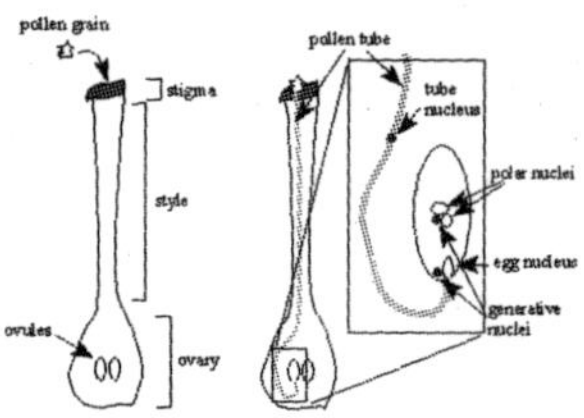

- **polliniferous**
  bearing pollen.

- **pollinium**
  pl. Pollinia Coherent masses or 'packets' of pollen. Orchids have two, four, six, or eight pollinium (packets). The number of pollinia is traditionally considered one of the major factors in defining a genus of an orchid.

- **polyad**
  1. a set of molecular states connected by vibratory resonance.

2. a patented name for the adsorption process for the removal and destruction of Volatile Organic Compounds (VOC).

- **PHB**
  Poly-Beta-Hydroxybutyrate. Common storage material of prokaryotic cells consisting of beta-hydroxybutyrate or other beta-alkanoic acids.

- **polybrid**
  a hybrid with more than two parental groups.

- **polycarpellary**
  describes a fruit which consists of two or more fused carpel.

- **polyclonal antiserum**
  mixture of antibodies to a variety of antigens or to a variety of determinants on a single antigen.

- **polycyclic**
  of a disease or pathogen: Producing many generations of innoculum and many cycles of infection during a single growing season.

- **polyembryony**
  the condition in which a seed has more than one embryo.

- **polyethylene**
  a plastic which allows the passage of gases but not of moisture.

- **polyetic**
  of plant disease epidemics: continuing from one growing season to the next.

- **polygamodioecious**
  describes a plant group which has bisexual and male flowers on some plants, and bisexual and female flowers on others.

- **polygamous**
  1. bearing male and female flowers on the same plant.
  2. having antheridia and archegonia disposed in various ways on the same plant.

- **polygenic**
  a character controlled by many genes.

- **polymer**
  large molecule formed by polymerisation of monomeric units.

- **polymerase chain reaction**
  the selective amplification of DNA by repeated cycles of heat denaturation of the DNA, annealing of two oligonucleotide primers that flank the DNA seg-

ment to be amplified and the extension of the annealed primers with the heat insensitive Tag DNA polymerase.

- **polymorphic**
  having more than two distinct morphological variants.

- **polymorphism**
  pleomorphism.

- **polynucleate**
  having more than one nucleus per cell.

- **polypetalous**
  having separate petals.

- **polyphyletic**
  having members that originated, independently, from more than one evolutionary line.

- **polyploidy**
  having more than two sets of chromosomes in cells of the sporophyte or more than one set in cells of the gametophyte.

- **polypore**
  a member of the Polyporaceae, the hymenium forms tubes in the basidiocarp.

- **polysaccharide**
  long chain of monosaccharides (sugars) linked by glycosidic bonds.

- **polysepalous**
  having a calyx made up of separate sepals.

- **polysome**
  strings of ribosomes attached by strands of mRNA.

- **polystichous**
  having leaves or other structures arranged in distinct rows.

- **polytrichous**
  covered with many hairs.

- **polytypic**
  containing more than one taxon of the next lower rank, e.g., a genus with more than one species.

- **pome**
  a kind of fleshy fruit, particularly an apple.

Pome - fruit within hypanthium

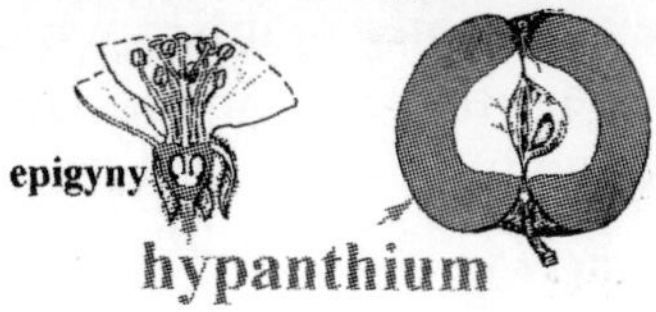

- **pompon**
  a small globular flower or flower head.

- **pond**
  a small, still body of water, an artificial body of water in a garden usually containing aquatic plants and fish.

- **pondweeds**
  a popular name for species of wetland plants that grow partially or wholly underwater.

- **pool**
  a small and relatively deep body of usually fresh water, either standing or as part of a stream. There is little wave action and often a reflective quality.

- **population**
  plants of a species growing in a given place at a given time. Two similar species may mingle, e.g., *Solidago canadensis* and *Solidago altissima*, but the botanical term 'population' would include one species or the other, not both.

- **population density**
  the concentration of individuals in relation to the space they occupy, how close individuals occur.

- **porate**
  describes a pollen grain which has rounded apertures only.

- **pore**
  a very small opening that is usually circular.

- **pore space**
  portion of soil bulk volume occupied by soil pores.

- **poricidal**
  a type of dehiscence in which the pollen is released through pores at the tip of the anther.

- **porin**
  a protein channel in the lipopolysaccharide layer of Gram-negative bacteria.

- **poroconidium**
  see **porospore**.

- **porose**
  pierced with small holes or pores.

- **porosity**
  volume of pores in a soil sample (non-solid volume) divided by the bulk volume of the sample.

- **porospore**
  a conidium produced by an extension of the inner wall of the conidiogenous cell and extrusion through a pore in the wall of the conidiophore.

- **porrect**
  spreading outward and forward.

- **portland cement**
  cement consisting. Predominantly of calcium silicates which reacts with water to form a hard mass.

- **positive drainage**
  surface or underground pipes or trenches to remove excess water.

- **post-emergent herbicide**
  a pesticide that kills plants after they have grown to seedling stage or beyond.

- **posterior**
  in an axillary flower, on the side nearest to the axis of inflorescence.

- **postreproductive**
  refers to an organism which is past the age of breeding capacity.

- **potager**
  giving credit to the French who inspired this vegetable garden. It is planted in a formal and ornamental style.

- **potato fork**
  a fork of heavy construction with flat tines, for deep digging as is needed for potato harvesting.

- **potbound**
  the condition of a houseplant or outdoor container plant which has been left in the same pot or container for too long a period of time and the roots fill every niche of the pot or container. A hook or blade is required to cut through and separate the tangled, matted outer roots when removed from the pot and before placing in the new container or the plant may never recover from this disastrous condition.

- **potential hydrogen**
  a log scale measurement of the acidity/alkalinity of a solution with 1 being ex-

tremely acidic, 10 being extremely alkaline, and 7 being neutral. Most plants prefer a soil within a certain range of pH.

• **potherb**
a plant whose leaves, stems, or flowers are cooked or used as a seasoning.

• **pothole**
1. a pond or pool utilised by waterfowl for nesting and raising young.
2. a hole worn in solid rock by the action of water, especially when a pebble repeatedly spins in the depression.

• **potpourri**
a mixture of sweet smelling leaves, petals, blooms to create a perfume in a room. Tussie massies in the Victorian era were popular nosegay potpourri creations.

• **potting medium**
material used for growing plants in containers. Mixes may include vermiculite, perlite, sand, peat, charcoal, loam and fertiliser.

• **potting on**
the re-potting of a plant into a proper-sized larger pot which will allow continued root development.

• **potting up**
taking the young seedlings or transplants into a specific container for mature growth.

• **potyvirus**
siglum of potato virus Y. Member of a large group of plant viruses with flexuous particles containing a single molecule of linear RNA, most of which are transmitted by aphids in a non-circulative manner.

• **poultice**
moist, usually hot, mass of plant material, wrapped in a cloth and applied to the skin to bring about some desired action, such as bringing a boil to a head.

• **pour plate**
method for performing a plate count of micro-organisms. A known amount of a serial dilution is placed

in a sterile Petri dish and then a melted agar medium is added and the innoculum mixed well by gently swirling. After growth the number of colony forming units is counted.

- **powdery mildew**
  a fungus forming a white powdery coating on leaves and stems.

- **power take off**
  unit on a tractor to attach power tools.

- **pozzolan cement**
  volcanic rock powdered and used in making hydraulic cement.

- **pradines**
  lopping tools.

- **praemorse**
  appearing as if bitten off.

- **prebloom**
  the stage or period immediately preceding blooming.

- **precleaning**
  the process for removing the bulk of foreign materials grossly different in size from the crop seed.

- **precocious**
  occurring early, as flowers appearing before the leaves, hysteranthous.

- **predation**
  relationship between two organisms whereby one organism (predator) engulfs or captures and digests the second organism (prey).

- **predisposition**
  an increase in susceptibility resulting from the influence of environment on the suscept.

- **pre-emergent**
  an herbicide, or combination of fertiliser and herbicide, that is applied to a lawn early in the season, before the new growth has occurred.

- **pre-emergent herbicide**
  a pesticide that kills plants as they germinate.

- **pre-emergent weed killer**
  a great idea in good gardening. Although not organic, this is using a herbicide to kill the weed seeds to prevent them from germinating.

- **preference-follower grazing**
usage: Not a recommended term.

- **pre-reproductive**
refers to a plant or animal which has not yet matured sexually.

- **prescribed burning**
the intentional burning of plant material in an area. Sometimes used to consume underbrush and other fuel and thus prevent larger, more destructive fires. Also used to maintain a stable prairie or grassland by aborting the process of succession to shrubs and trees.

- **preservative**
an additive used to protect against decay, discolouration, or spoilage.

- **press straps**
bands or ropes to hold the plant press tightly together. Traditionally made of web and buckles, but now often with Velcro.

- **pressure treated lumber**
lumber that has been treated with chemicals to prevent rotting.

- **prevernal**
appearing in late winter or very early spring, as do crocus flowers.

- **pricking out**
1. moving seedlings from a seed tray and replanting them with more space for development.
2. the moving of seedlings from the tray or pot in which they were sown to other receptacles where they can be spaced out individually.

- **prickle**
a small, sharp outgrowth involving only the outer epidermal layer, usually more slender than a thorn. This is the correct term for rose thorns.

Desert Prickle

- **primary cycle**
of plant disease: the first infection cycle to begin in a given season, usually occurring only once per season.

- **primary forest**
  a tract of trees with minimal impact from humans.

- **primary infection**
  the first infection of a plant by a pathogen emerging from a dormant stage in its life cycle (overwintering or oversummering).

- **primary inoculum**
  the over wintering or over summering pathogen or its propagules that cause primary infection.

- **primary producer**
  organism that adds biomass to the ecosystem by synthesising organic molecules from carbon dioxide and simple inorganic nutrients.

- **primer**
  molecule (usually a polynucleotide) to which DNA polymerase can attach the first nucleotide during DNA replication.

- **primocane**
  the floricane, but in the first year of growth before it is capable of flowering.

- **primordial utricle**
  the layer of protoplasm lying next to the cell wall, which is very often conspicuous when dried and shrunken.

- **primordium**
  a tissue or organ in the early stages when it has undergone differentiation changes but has not yet matured enough to emerge.

- **priority**
  the rule of nomenclature whereby the first published of two validly published names for the same entity is given approval as the accepted name.

- **prismatic**
  of the shape of a prism, angular, with flat sides, and of nearly uniform size throughout.

- **pro hybrida**
  means 'as a hybrid.' Used in citations to show that a plant now known as a species started as a hybrid.

- **pro parte**
  means 'in part.' Used in citations to show that a taxon as used by one author shows only a portion of what was intended by the original author.

- **pro specie**
  means 'as a species.' Used in citations to show that a plant, which had been accepted as a species is now designated as a hybrid.

- **probasidium**
  the cell in which karyogamy occurs in the basidiomycetes. (pl. probasidia)

- **probe**
  a specific sequence of DNA or RNA used to detect complementary sequences by hybridisation.

- **proboscis**
  beak.

- **procorpus**
  in nematodes: cylindrical portion of the corpus anterior to the metacorpus.

- **procumbent**
  lying on the ground or trailing but without rooting at the nodes.

- **prodelphic**
  in nematodes: having uteri parallel and anteriorly directed at the origin.

- **profundal zone**
  the area of deep water at the bottom of a lake below effective penetration of light.

- **progeny**
  plants grown from seeds produced by parent plants, offsprings.

- **projecting**
  refers to stamens, styles, and stigmas which exert beyond the petals of a flower.

- **prokaryote**
  organism lacking a unit membrane-bound nucleus and other organelles, usually having its DNA in a single circular molecule.

- **prolepsis**
  a hyperplastic symptom of disease in which organs appear before the natural time (pl. prolepses).

- **proliferation**
  a rapid and repeated production of new cells, tissues, or organs, specifically, a hyperplastic symptom of plant disease in which organs continue to develop after they have reached the point beyond which they normally do not grow.

- **proliferous**
  1. free producing offshoots, bulblets, or plantlets.

2. in mosses, bearing young shoots from the antheridial or archegonial cluster of leaves.

- **prominent**
standing out from the surface, clearly visible and touchable, as are veins and other surface features.

- **prominulous**
slightly raised above the adjoining tissue.

- **promoter**
a region of DNA, usually upstream of a coding sequence, that binds RNA polymerase and directs the enzyme to the correct transcriptional start site.

- **promycelium**
the short hypha bearing sporidia produced by the teliospore, the basidium. (pl. promycelia).

- **propagation**
in gardening usage, this refers to the many different ways of starting new plants.

- **propagative virus**
a circulative virus that replicates in its insect vector. Such a virus is said to be Propagatively Transmitted (e.g., potato yellow dwarf virus).

- **propagule**
any structure having the capacity to give rise to a new plant, whether through sexual or asexual (vegetative) reproduction. This includes seeds, spores, and any part of the vegetative body capable of independent growth if detached from the parent.

- **prophage**
state of the genome of a temperate virus when it is replicating in synchrony with that of the host, typically integrated into the host genome.

- **prophyll**
1. the bracteole under a single flower or pedicel.
2. the showy first bract borne on the peduncle and, occasionally, on the inflorescence branches of some monocots.

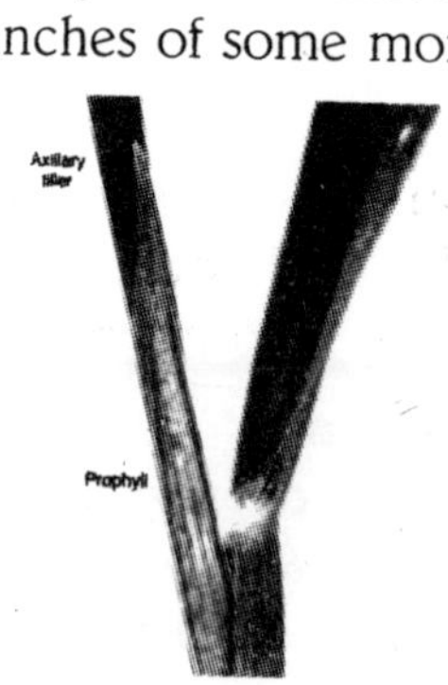

• **prosenchymatous**
cells with pointed ends dovetailing into each other.

• **prosthetic group**
tightly bound, non-protein portion of an enzyme, not the same as coenzyme.

• **prostrate**
lying flat upon the ground.

• **protandrous**
refers to a flower where the shedding of the pollen occurs before the stigma is receptive.

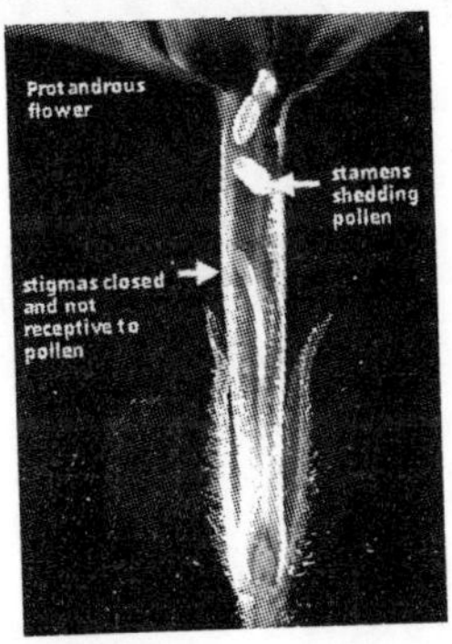

• **protease**
a generic term for an enzyme that cleaves a polypeptide chain

• **protectant**
any chemical agent that interacts with a pathogen on the plant surface to inhibit infection before it takes place. Non-systemic.

• **protected area**
a legally restricted land or water area under public or private ownership, which has managed to achieve specific conservation objectives.

• **protection**
a principle of plant disease control in which a barrier is placed between suscept and pathogen (e.g., the use of protective chemical dusts or sprays)

• **protein fraction**
solubility of proteins, as originally termed by Osborne, who designed a sequential extraction scheme based on water, dilute salt, 70% (v/v) alcohol, and dilute acid or alkali.

• **protein quality**
1. balance of essential amino acids in protein.
2. biological availability of protein. In general, for cereals the first limiting amino acid and the one that affects protein quality the most is lysine.

• **proteinaceous**
of, relating to, resembling, or being protein.

• **proterogynous**
having the stigma ripe for the pollen before the maturity of the anthers of the same flower.

• **prothallium**
the minute scale-like growth from the spore of a fern.

• **prothallus**
a cellular, usually flat and thallus-like growth, resulting from the germination of a spore, upon which are developed sexual organs or new plants. The gametophyte stage of ferns.

• **protista**
old taxonomic term referring to algae, fungi, and protozoa (collectively, the eukaryotic protists), and the prokaryotes.

• **protocooperation**
see **synergism.**

• **protocorn**
the first growth produced by a germinating orchid seed before the growth of leaves.

• **protogynous**
refers to a flower where the shedding of the pollen occurs after the stigma has ceased to be receptive.

• **protologue**
refers to all parts of botanical publication, including, but not limited to, diagnosis, description, synonymy, citation of specimens, illustrations, etc., expressed as data, information, and knowledge.

• **PMF**
Proton Motive Force. Energized state of a membrane created by expulsion of protons through action of an electron transport chain.

• **protonema**
the green, branched, alga-like threads produced from the spore and often persis-

tent during the lifetime of the plant produced from it.

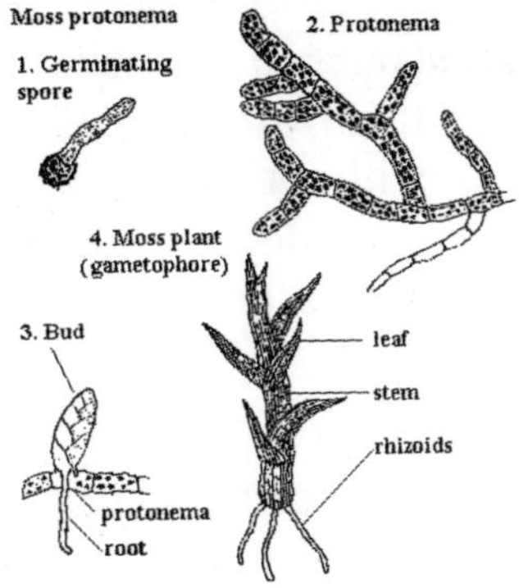

- **protoplasm**
  complete cellular contents, cytoplasmic membrane, cytoplasm, and nucleus, usually considered the living portion of the cell, thus excluding those layers peripheral to the cell membrane.

- **protoplast**
  a plant cell from which the cell wall has been removed.

- **prototroph**
  a strain of micro-organism whose nutritional requirements do not exceed those of the corresponding wild-type strain

- **protoxylem**
  the first formed xylem.

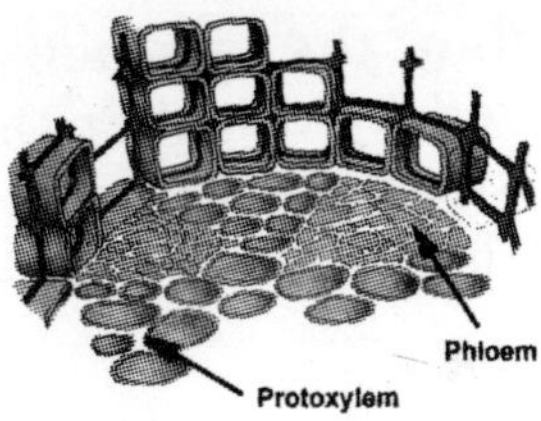

- **protozoan**
  unicellular eukaryotic micro-organisms that move by either protoplasmic flow (amoebae), flagella (flagellates) or, cilia (ciliates). Most species feed on bacteria, fungi, or detrital particles.

- **provirus**
  see **prophage**.

- **proximal**
  toward the base.

- **proximate**
  close together.

- **proximate analysis**
  analytical system that includes the determination of ash, crude fibre, crude protein, ether extract, moisture (dry matter), and nitrogen-free extract.

- **pruina**
  a woolly white covering of the upper cortex of some lichens.

- **pruinose**
  having the surface obscured by a rather thick, bluish or greyish, wax-like coating.

- **pruning**
  the cutting off leaves or branches within limits in order to remove dead or diseased foliage or branches. Also used to control or direct growth, increase quality or yield of flowers or fruit, and to ensure growth position of main branches to enhance structural strength.

- **pruning saw**
  a serrated blade tool for cutting small limbs, may be styled so the blade folds into the handle, or fit into a sheath.

- **pruning shears**
  specialised scissors to cut plants back.

- **pseud**
  a prefix meaning 'apparently but not actually.' False.

- **pseudanthium**
  an inflorescence which looks like a simple flower, but is made of more than one axis with a number of flowers.

- **pseudobulb**
  thickened or bulb-like stems (called 'pseudobulbs' because they are not true bulbs) produced by some sympodial orchids to store water and food. Only orchids whose habitat has seasonal periods of dryness or drought have adopted this life-saving characteristic.

- **pseudocarp**
  a fruit, such as the strawberry or pear, that develops not only from the ripened ovary, or ovaries, but from non-ovarian tissue as well.

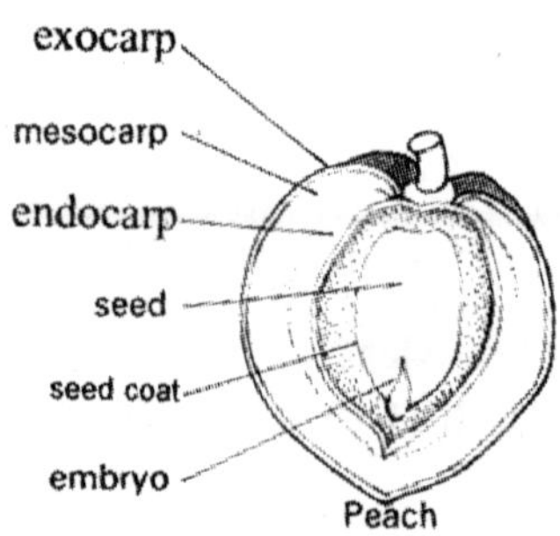

• **pseudocoel**
in nematodes: body cavity containing a fluid in which the various internal organs are suspended.

• **pseudocyphellae**
pores found in the upper or lower cortex of lichens.

• **pseudolamina**
a false lamina, one usually formed by the lateral expansion of a rachis.

• **pseudomonad**
member of the genus Pseudomonas, a large group of Gram-negative, obligately respiratory (never fermentative) bacteria.

• **pseudoparaphysis**
a sterile thread that grows downward in the cavity of some ascocarps, usually becoming attached at the bottom.

• **pseudoparenchyma**
filamentous tissue compacted to resemble parenchyma.

• **pseudopodetium**
an upright, fruticose thallus of some lichens.

• **pseudopodium**
protrusion of an amoeboid cell formed by the extrusion or streaming of the cytoplasm (but still enclosed in the membrane) for the purpose of movement or feeding.

• **pseudopods**
literally, false feet, a temporary protrusion of the protoplasm of a cell for the purpose of moving or feeding.

• **pseudostem**
an erect aerial growth which appears to be a stem with leaves, but is actually packed or overlapping sheaths and stalks of essentially basal leaves.

• **pseudoterminal**
1. describes a bud, which is situated at what appears to

be the apex of a branch, but is in fact located in an axillary position beside a leaf scar.
2. refers to a bud, which replaces an injured terminal bud which has died.

- **pseudothatch**
the upper surface layer above a thatch, made up of relatively non-decomposed leaf remnants and clippings.

- **pseudothecium**
an ascostroma resembling a flask-shaped perithecium. (pl. pseudothecia).

- **psychrophile**
organism able to grow at low temperatures and showing a growth temperature optimum at less than 15° C.

- **pteridology**
the study of ferns.

- **pteridophyte**
any of the non-flowering vascular plants of the division Pteridophyta having roots, stems, and leaves, a fern or fern ally.

- **pterophyta**
a major taxonomic division containing the ferns, showing clear alternation of generations with a dominant vascular sporophyte which begins with the very reduced gametophyte.

- **puberulose**
minutely pubescent.

- **pubescence**
a covering of soft, short hairs.

- **pubescent**
covered with hairs, especially if short, soft, and down-like.

- **puddle**
1. a very small pool of water, often transitory, appearing during heavy rain and disappearing shortly afterward.
2. to knead or work soil while wet, making it impervious to water.
3. to work while wet, as in flooded rice fields.

- **puffiness**
sponge like condition of turf, resulting in an irregular surface.

- **pulaski clearing axe**
a tool used mostly by fire fighters that is a combination of axe and grub hoe.

• **pulp**
1. the soft, succulent part of a fruit, usually composed of mesocarp.
2. the pith of a stem.

• **pulpwood**
the wood of trees like aspen, populus, and spruce, picea, which is easily ground into a soft mass for the production of paper

• **pulverulent**
powdered, appearing as if covered by minute grains of dust.

• **pulvinate**
like a cushion.

• **pulvinus**
a swelling at the base of the stalk of a leaf or leaflet.

• **punctate**
dotted with depressions or with translucent internal glands or coloured dots

• **puncticulate**
minutely punctate.

• **pungent**
terminating in a rigid sharp point. Acrid to smell or taste.

• **pupa**
a stage in complete metamorphosis when an insect transforms from the larval to adult stage of development.

• **pure culture**
population of micro-organisms composed of a single strain. Such cultures are obtained through selective laboratory procedures and are rarely found in a natural environment.

• **pure live seed**
percentage of pure germinating seed: (pure seed percentage x germination percentage)/100.

• **purgatives**
medicines which empty the intestines.

• **pustular**
like a blister.

• **pustulate**
covered with blisters

• **pustule**
small blister-like elevation of the leaf epidermis created as

spores emerge from underneath and push outward.

• **pustuliform**
blister-like.

• **putamen**
the shell of a nut, the bony part of a stone fruit.

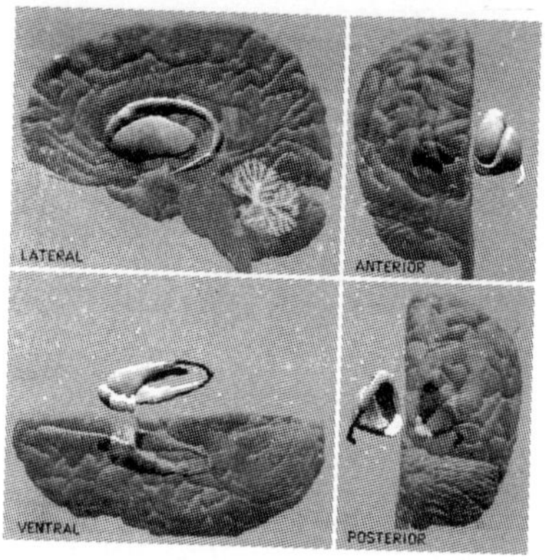

• **put-and-take stocking**
the use of variable animal numbers during a grazing period or grazing season, with a periodic adjustment in animal numbers in an attempt to maintain desired management criteria (e.g., a desired quantity of forage, degree of defoliation, or grazing pressure).

• **PVC sleeves**
a plastic pipe used for moving water that acts as a conduit.

• **pycnidiospore**
a conidium formed in a pycnidium.

• **pycnidium**
a closed sporocarp, usually opening by a pore, that contains a cavity bearing conidia.

• **pycniospore**
a spore (spermatium) borne in a pycnium in the Uredinales.

• **pycnium**
in the Uredinales, stage 0, consisting of the male fertilising elements (pycniospores) and female elements, (the flexuous hyphae).

Pycnidium

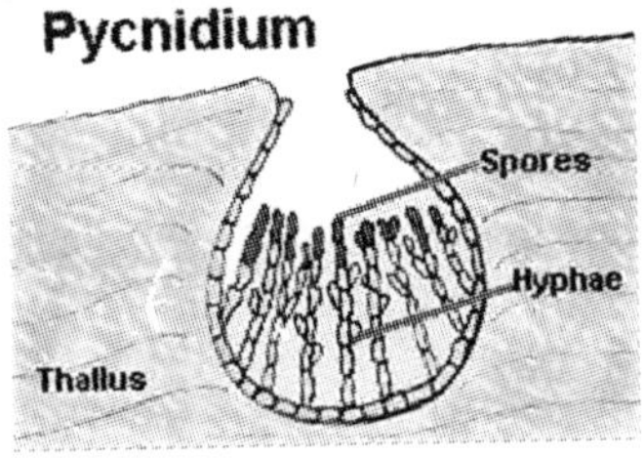

• **pycnospore**
an obsolete term for a pycnidiospore or a pycniospore.

• **pygmism**
the state of being dwarfed or reduced in size.

- **pyramidal**
  conical with flat angular sides.

- **pyrene**
  the stone of a succulent fruit made up of the endocarp and the seed.

- **pyrenomycete**
  one of the Pyrenomycetes.

- **pyrenomycetes**
  class of Ascomycotina, traditionally based on taxon with perithecioid ascomata that are ascohymenial in ontogeny and have unitunicate asci, often with an apical annulus

- **pyriform**
  pear-shaped.

- **pyxis**
  a capsule which opens in a circumscissile manner.

- **quadrangular**
  shaped into a square cross section, as are the stems of most mints, e.g., Labiatae.

- **quadrate**
  describes cells that are square or nearly so.

- **quadrigeneric**
  pertaining to four genera, used particularly in reference to hybrids combining members of four genera.

- **quadrinary combination**
  binomial plus two more taxon names such as subspecies and variety.

- **quadripinnate**
  four times pinnate.

- **quality control**
  the use of any device, technique, method, or person that makes possible the production, on time and at the lowest cost, of products of the quality necessary to gain full customer acceptance.

- **quarantine**
  legal restriction of the movement of plant pests (or the products) that may be harbouring them into areas where they do not occur.

- **quartered**
  having the petals formed into a number of sections, usually four but sometimes three or five.

- **quiescence**
  the absence of growth, usually inferring the absence of environmental conditions favouring growth. Usage: Although dormant seed are

quiescent, quiescence is distinguished from dormancy, which implies the inability to germinate even in the presence of environmental conditions favouring growth.

- **quinate**

having five leaflets growing from the same point of attachment.

- **quincunx**

the arrangement of five objects, such as trees, with one on each of four corners and one in the centre.

- **race**

a phenotypically, geographically, ecologically, and/or genetically distinct group, usually within a subspecies. Usage: Not covered by the botanical code of nomenclature. Compare cultivar, strain, variety.

- **raceme**

a simple unbranched infloresence in which the elongated axis bears flowers on short stems (pedicels) succession toward the apex.

- **racemiform**

refers to an inflorescence with the outward appearance of a raceme.

- **racemose**

resembling a raceme.

- **rachilla**

the axis of a spikelet, particularly in grasses.

- **rachis**

1. the main stalk of a flower cluster or the main leafstalk of a compound leaf.
2. in ferns, the continuation of the stipe through a compound frond.

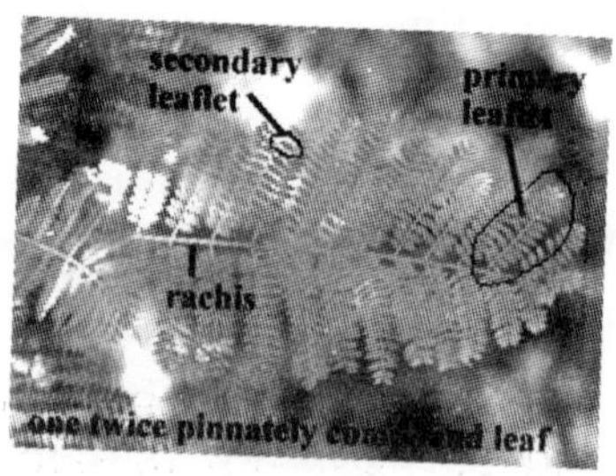

• **radial**
in flowers, one with the symmetry of a symmetry wheel

• **radiate**
spreading from or arranged around a common centre, bearing ray flowers.

• **radical**
belonging to or proceeding from the root or base of the stem near the ground.

• **radicle**
the portion of the embryo below the cotyledons that will form the roots, more properly called the caudicle.

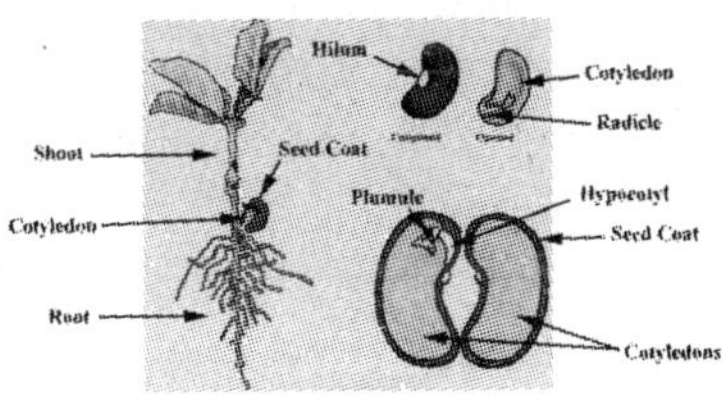

Cotyledons

• **radicums**
these plants are special in that their stems have roots that will cling as they grow vertically or grow over the ground. Ivy is a good example.

• **radius**
1. the outermost flowers in a daisy-like head, Composite, ray flowers as opposed to disc flowers.
2. the distance from the centre of a circle to the outer edge, as from the heart of a tree to the bark.

• **raffia**
a fibre-like material used for tying grafts and other horticultural purposes.

• **raised beach**
an old beach terrace at an elevation above the present shoreline, representing a formerly higher lake level.

• **raised bed**
a gardening area where the soil has been elevated above ground level. This gardening technique is especially used where soil drainage is poor. Beds can be raised in a structure of wood, brick, cement blocks, etc.

• **ramble**
to grow freely, often over another plant or structure.

• **rameal**
belonging to a branch.

• **ramentaceous**
refers to stems or leaves which have small, loose, brownish scales.

- **ramet**

1. an individual member of a clone.
2. an underground tree system giving rise to large suckering colonies, like bigtooth aspen trees, Populus grandentata.

- **ramification**

branching.

- **ramiflorous**

bearing flowers directly from large branches and leafless twigs, but not on the trunk.

- **ramiform**

with branches, branch-like.

- **ramose**

having many branches.

- **ramuli**

minute branchlets.

- **range**

of a plant pathogen: the geographical region or regions in which it is known to occur

- **range management**

the science of maintaining maximum range forage production without jeopardy to other resources or uses of the land.

- **range of variation**

the extremes of possibilities which may occur within a species and still remain a member of that species, including such things as leaf shape, leaf margins, flower colour, height, etc.

- **rank**

1. a vertical row, as of leaves. When you sight along the length of a branch from the tip end, if it appears there are two rows of leaves, either opposite or alternate, the branch is 2-ranked, if three rows, it is 3-ranked, etc.
2. in taxonomy, the position of a taxon in the hierarchy, e.g., species, genus, family,

- **raphe**

the part of the funiculus that is fused along the side of the ovule.

- **raphides**

needle-like crystals of calcium oxalate that occur in bundles in some plant cells.

- **rasping**
mouthparts that are rough and used to scrape a surface to feed.

- **ratchet tool**
one which uses mechanical means to multiply cutting or turning power, reducing cramping of hands and forearms.

- **ration**
the total amount of feed (diet) allotted to one animal for a 24-hour period.

- **ravine**
a depression worn by running water, larger than a gully and smaller than a valley.

- **raw cotton**
ginned lint that has not been subjected to any textile manufacturing process.

- **ray**
the branch of an umbel, the marginal flowers of an inflorescence when distinct from the disk.

- **ray floret**
one of the broad, conspicuously coloured florets of a compound flower, the structure of which suggests a single petal of an ordinary flower. These form the radiating border in the open-cantered type dahlias, or massed together, the flower head in fully double types.

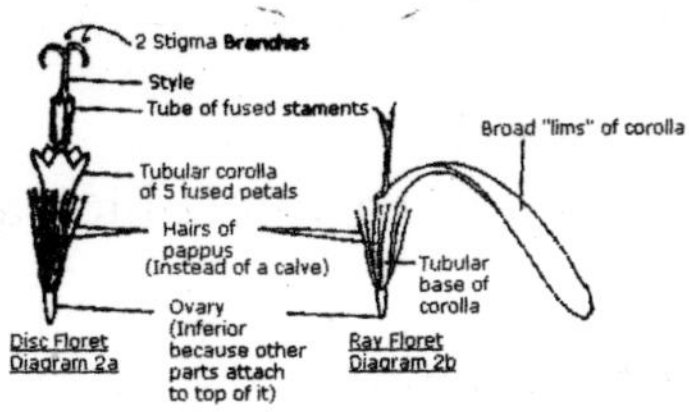

- **ray flower**
the bilaterally symmetrical flowers around the edge of the head in many members of Composite, each ray flower resembles a single petal.

- **rays**
radial strands of cells in wood and phloem.

- **realised genetic gain**
observed difference between the mean phenotypic value of the offspring of the selected parents and the phenotypic value of the parental generation before selection.

- **reannealing**
process where two complementary single strands of DNA automatically hybridise back into a single,

double-stranded molecule upon cooling.

- **rebuilding**
  the practices that result in complete change of a turf area.

- **recalcitrant**
  resistant to microbial attack.

- **recalcitrant seed**
  seed that does not survive drying and freezing.

- **receptacle**
  1. the more or less expanded or produced portion of an axis which bears the organs of a flower (the torus) or the collected flowers of a head, and in roses, enfolds the developing ovaries to form a hip.
  2. any similar structure in cryptogams.

Parts of a Typical Flower

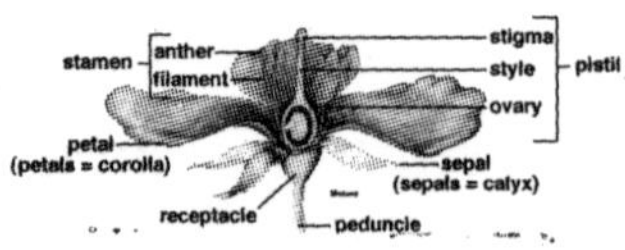

- **receptacular**
  carried on the receptacle, pertaining to the receptacle.

- **recession**
  a drop in the water line of a body of water, with a corresponding exposure of the bottom.

- **reclinate**
  bent or turned downward.

- **reclining**
  the lower portion somewhat flattened along the ground but the upper parts curving upward.

- **recombinant DNA**
  DNA molecules in which sequences, not normally contiguous, have been placed next to each other by in vitro methods.

- **recombination**
  process by which genetic elements in two separate genomes are brought together in one unit.

- **recropping**
  a system in which land is cropped annually without a fallow period (i.e., continuously cropped year after year), or in a cropping rotation longer than one year. The land may be in a cereal grain, or an alternate crop may be inserted into the rotation.

- **rectum**
  in nematodes: posterior gut of the female. A narrow, dor-

soventrally flattened tube that is lined with cuticle and separated from the intestine by a sphincter muscle.

- **recurved**
  curved downward or backward, with ray florets curved backward along their length toward the stem.

- **reduced tillage**
  a tillage system with aggressive primary tillage, may be performed in conjunction with special planting procedures to reduce or eliminate secondary tillage operations.

- **reducing equivalent (power)**
  electrons stored in reduced electron carriers such as NADH, NADPH and $FADH_2$.

- **reduction**
  process by which a compound accepts electrons.

- **reduction potential**
  inherent tendency of a compound to act as an electron donor or an electron acceptor. Measured in milli-volts.

- **reeds**
  tall thin wetland plants of the genera Phragmites and Sparganium, a term often incorrectly applied to bulrushes.

- **reel mower**
  a mower with multiple blades mounted on a cylinder. The blades cut against a bar. It makes precise cuts and is ideal for lower mowing heights. The blades require professional sharpening. These mowers are safer to use than rotary mowers.

- **reference**
  a citation to a source of information, such as a written document.

- **reflexed**
  abruptly bent or turned downward.

- **refrigerants**
  herbal preparations which cool the body.

- **registered seed**
  progeny of breeder, select, or foundation seed handled

under procedures acceptable to the certifying agency to maintain satisfactory genetic purity and identity.

- **regular**
  uniform in shape or structure.

- **regular flower**
  generally symmetrical and uniform in the number of its parts.

- **regulation**
  the degree to which a limiting factor or similar element actually controls a process.

- **rehabilitation**
  the process of improving specific ecosystem factors in a degraded habitat.

- **rejuvenation pruning**
  the practice of cutting all the main stems of a shrub back to within half-inch of the ground during winter dormancy.

- **relative humidity**
  the ratio of the quantity of water vapour actually present in the air to the greatest amount possible at a given temperature.

- **release rate, fertiliser**
  the rate of nutrient release following fertiliser application. Water-soluble fertilisers are termed quick-release or fast-release, while insoluble or coated soluble fertilisers are referred to as slow-release or controlled-release.

- **reliction**
  the exposure of a lake's bottom by natural recession.

- **REMI**
  Restriction Enzyme Mediated Integration. A method of transformation that generates tagged mutations

- **remontant**
  repeat bloomer. Plants that bloom more than once a year. E.g. Tea, olive, daylilies, or bottle bush.

- **renewal pruning**
  the practice of cutting all the main stems of a shrub back to within half-inch of the ground during winter dormancy.

- **renewal spur**
  on grapevines, the cane pruned to one or two nodes on the cordon, becomes the fruiting cane the following year.

- **renovation**
  removing an old planting and putting in a new one or removing and replacing only part of a planting. In strawberry culture, this process involves removing the leaves of the plants and cultivating the aisle to reduce the width of the row of plants to no more than 15 inches.
- **repand**
  with a slightly uneven and somewhat sinuate margin.
- **repellent**
  a substance used to keep insects or animals from an area.
- **repent**
  creeping, and typically rooting at the joints.
- **replicates**
  duplicate specimens.
- **replicon**
  any DNA sequence or molecule that possesses a replication origin and is therefore potentially capable of being replicated in a suitable cell.
- **replum**
  the hardened framework of the placenta which remains when the outer walls or valves of a silique fall away, the septum which holds the seeds.

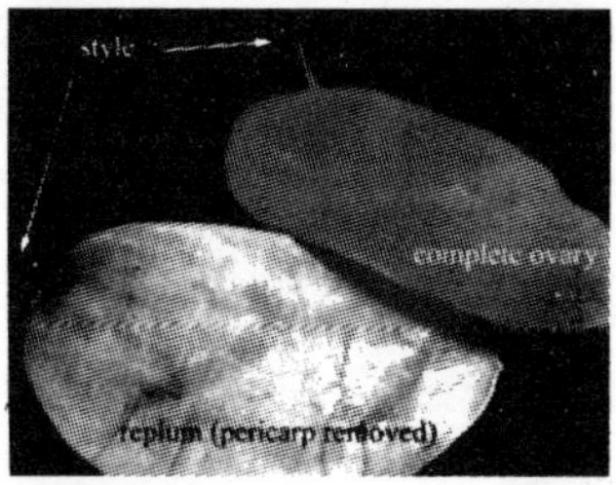

- **repression**
  process by which the synthesis of an enzyme is inhibited by the presence of an external substance (the repressor).
- **re-seeding**
  plants that drop their seeds for next season. Called 'easily re-seeders'. Wildflowers and weeds are the biggest categories.
- **residual response, fertiliser**
  the delayed or continued turf grass response to slow-release fertilisers, lasting longer than the usual initial response from water-soluble (i.e., quick-release) fertilisers.
- **residue biomass**
  the biomass that remains after removal or utilisation

of part of the biomass by grazing, harvesting, burning, etc.

• **resin dot**
tiny circular or globular yellow spots, usually not obvious except under magnification

• **resiniferous**
producing resin.

• **resins**
sticky-to-brittle plant products derived from essential oils and sometimes possessing marked odours, used in medicines, varnishes, etc.

• **resistance**
the ability of an organism to exclude or overcome, completely or in some degree, the effect of a pathogen or other damaging factor.

• **respiration**
catabolic reactions producing ATP in which either organic or inorganic compounds are primary electron donors and exogenous compounds are the ultimate electron acceptors.

• **response**
the change produced in an organism by a stimulus

• **resting period**
mostly in terms of bulbs. It is a period of dormancy where energy is restored to the plant.

• **resting spore**
a thick-walled spore, particularly one formed by a sexual process, germinating only after an extended period of dormancy (e.g., an overwintering teliospore).

• **restoration**
the return of an ecosystem or habitat to its original structure, natural complement of species, and natural functions.

• **restriction endonuclease**
1. an endonuclease that binds to double-stranded DNA at a specific nucleotide sequence and then, if both strands of the DNA lack appropriate modification at that sequence, cleaves the DNA either at the recognition sequence or at another site in the DNA molecule.
2. enzyme that recognises and cleaves specific DNA sequence, generating either blunt or single-stranded (sticky) ends.

- **rest-rotation grazing**
  usage: Not a recommended term.

- **resupine**
  refers to a flower or leaf on which the stalk twists a full half turn, 180 degrees, so that the organ appears to be upside down.

- **retaining wall**
  a wall that has been built on a slope to keep the soil from sliding or eroding. It would stabilise the area.

- **retardants**
  admixtures that increase the setting time of cement by slowing down hydration.

- **retentive sepals**
  sepals that remain attached to the apex of the receptacle after it has ripened into a hip.

- **reticulate**
  marked with a branched network of veins or fibres.

- **reticulum**
  a mesh-like network.

- **retinaculum**
  the structure by which another structure is attached to a plant.

- **retrorse**
  directed back or downward

- **retrovirus**
  virus containing single-stranded RNA as its genetic material and producing a complementary DNA by action of the enzyme reverse transcriptase.

- **retuse**
  with a shallow notch at a rounded apex.

- **reverse osmosis**
  a process used to purify water by forcing contaminated water through a semi-permeable membrane. The membrane allows the water molecules to pass through but not the other substances contaminating the water. Reverse Osmosis is used to commercially purify seawater as well as by hikers to remove impurities from water found along the trail.

- **reverse transcription**
  process of copying information found in RNA into DNA.

- **reversion**
  when a genetically differentiated plant or sport reverts to normal growth. For instance, variegated leafs turn green.

- **revert**
  sometimes a particular cultivar might change back to one of its original species. e.g., Variegation going to solid green, some doubles going to single.
- **revolute**
  rolled backward from the margins or apex. Describing a ray floret with margins rolled backward along their longitudinal axis.
- **RFLP**
  Restriction Fragment Length Polymorphism. Inherited differences in sites for restriction enzymes (e.g., caused by base changes in the target site that result in differences in the lengths of the fragments produced by cleavage with the relevant restriction enzyme.) RFLPs are used for genetic mapping to link the genome directly to a conventional genetic marker
- **rhabdions**
  in nematodes: plates in the cuticular lining of the stoma that make up the walls of the various divisions of the stoma.
- **rhachilla**
  a secondary axis, specifically, in the grasses and sedges, the floral axis as opposed to that of the spike or spikelet.
- **rhachis**
  the axis of a spike or of a compound leaf.
- **rhape**
  the ridge or adnate funicle which in an anatropous ovule connects the two ends.
- **rhaphides**
  needle-shaped crystals often occurring in bundles within the cells of certain plants.
- **rhizines**
  strands of hyphae found on the lower surface of many foliose lichens.
- **rhizobacteria**
  bacteria that aggressively colonise roots.
- **rhizobia**
  bacteria capable of living symbiotically in roots of leguminous plants, from which they receive energy and often fix molecular dinitrogen.

• **rhizoid**
a root-like structure forming part of the thallus in certain algae and fungi. It may anchor the organism to the substratum and/or act as an absorptive organ.

• **rhizomatous**
having or appearing to have rhizomes.

• **rhizome**
the woody parts of the rootstock at the base of the orchid which grows along or just under the surface of the ground or along host. New growth of sympodial orchids always begins at the end of the rhizome.

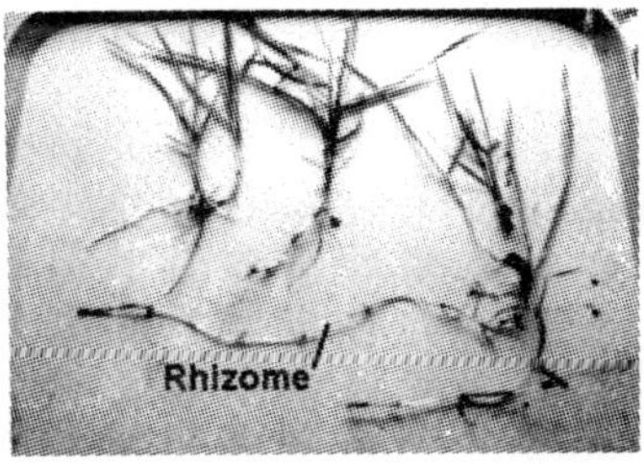

• **rhizome chaff**
small pieces of rhizomes which are spread upon the ground and covered with soil so they can take root and form new plants.

• **rhizome cutting**
a piece cut from a rhizome which can take root and form a new plant.

• **rhizomorph**
a macroscopic, typically rope-like strand of compacted tissue formed by certain higher fungi. Rhizomorphs often are enduring structures that can remain dormant under adverse conditions.

• **rhizomucor**
the Zygomycetous fungus is reported to be allergenic. It may cause mucorosis in immune compromised individuals. It occupies a biological niche similar to Mucor sp. It is often linked to occupational allergy. The sites of infection are the lung, nasal sinus, brain, eye and skin. Infection may have multiple sites.

• **rhizomycelium**
branching, anucleate or sparsely nucleate, rhizoidal filaments of variable width forming part of the thallus in some fungi.

- **rhizoplane**
plant root surfaces and usually strongly adhering soil particles.

- **rhizopus**
the Zygomycetous fungus is reported to be allergenic. It may cause mucorosis in immune compromised individuals. It occupies a biological niche similar to Mucor sp. It is often linked to occupational allergy. May cause mucorosis in immune compromised individuals. The sites of infection are the lung, nasal sinus, brain, eye and skin. Infection may have multiple sites.

- **rhizosphere**
an environment regarded, variously, as that region of the soil modified as a result of the uptake and deposition of substances by a growing root, the root itself, together with that volume of soil which it influences, the root surface together with that region of the surrounding soil in which the microbial population is affected by the presence of a root.

- **rhizosphere competence**
ability of an organism to colonise the rhizosphere.

- **rhodotorula**
a reddish yeast typically found in moist environments such as carpeting, cooling coils and drain pans. In some countries it is the most common yeast genus identified in indoor air. This yeast has been reported to be allergenic. Positive skin tests have been reported. It has colonised terminally ill patients.

- **rhomboid**
refers to leaves, tepals, etc., which are diamond-shaped, with the base and tip having acute angles and the sides having obtuse angles.

- **rib**
a primary or prominent vein of a leaf..

- **ribbed**
with one or more prominent veins or nerves.

- **ribonucleic acid (RNA)**
polymer of nucleotides connected via a phosphate-ribose backbone, involved in protein synthesis.

- **ribosomal RNA (rRNA)**
  types of RNA found in the ribosome, some participate actively in the process of protein synthesis.

- **ribs**
  geologic ridges running parallel, as often occurs along a mountain side.

- **rickettsiae**
  micro-organisms similar to bacteria in most respects but generally capable of multiplying only inside living host cells, parasitic or symbiotic.

- **ridged**
  angular, with lengthwise lines.

- **right of way**
  a strip of land across another person property, legally designated for owners of landlocked property to enter their holdings.

- **rind**
  1. the bark of a tree outside the cambium layer.
  2. the epicarp of a citrus fruit, e.g., an orange or a lemon.

- **ring scar**
  a ring-like scar left on a twig after a leaf drops.

- **ringed**
  having narrow encircling stipule scars around leaf scars.

- **ringent**
  gaping, said of lipped flowers with an open throat or mouth.

- **ringspot**
  a type of local lesion consisting of single or concentric rings of discolouration or necrosis, the regions between the concentric rings being green. The centre of the lesion may be chlorotic or necrotic.

- **riparian**
  of, or relating to, rivers or streams.

- **ripe**
  fully grown and developed, mature.

- **ripening**
  1. the maturing of a fruit as developmental changes prepare it for seed dispersal.
  2. the hardening of wood, especially of soft twigs, important in taking cuttings for propagation.

- **ripewood cutting**
  a cutting from ripened wood, usually used for evergreen plants.
- **RNA blotting**
  a technique for transferring RNA from an agarose gel to a nitrocellulose filter on which it can be hybridised to a complementary DNA .
- **rock garden**
  an area constructed of larger rocks arranged to look natural. Planted with plants that generally do not need a lot of care.
- **rockwool**
  this really is the state of the art rooting medium, and is used in plant propagation. It is alkaline and one will have to compensate for this.
- **rogue**
  a variation from the standard varietal type, so remove such undesirable plants (especially those infected with viruses) from the growing crop
- **roguing**
  the removal of diseased plants from a crop in order to prevent the spread of the disease.
- **rolled**
  curled under, revolute.
- **root**
  the portion of the plant usually found below ground. They are distinguished from stems by not having nodes.
- **root ball**
  matted roots plus enclosed soil within the pot of a container grown plant or when plant material is transplanted.
- **root crops**
  any vegetable of which the roots are edible: e.g., carrots, potatoes, turnips.
- **root culture**
  the in vitro growth of roots. Root culture usually refers to root tip (meristem) on a synthetic medium.
- **root cutting**
  the root is used for propagation. Plants that grow away from the mother plant and root can be dug up and transplanted.
- **root girdling**
  encircling roots at or below the surface of the ground that tend to strangle the plant.

- **root hairs**
tubular outgrowths of surface cells of the root.

- **root knot**
a disease of the roots characterised by swelling and caused by nematodes.

- **root nodule**
specialised structure occurring on roots, especially of leguminous plants, in which bacteria fix dinitrogen and make it available for the plant.

- **root pruning**
1. pruning to stimulate the growth of new roots within a parameter in preparation for transplanting.
2. trimming and/or scoring the outer layer of roots of a plant that has become root bound.
3. the act of removing a portion of a plant's roots to keep top growth in check.

- **root rot**
quite common in plants that are affected by fungus diseases and have poor drainage.

- **root zone**
the entire area where roots are growing below the plant. Root zones are important for walking paths and future plantings.

- **rootball**
the root stock and surrounding soil together, in nurseries and during transport, held together by burlap or other wrapping.

- **rootbound**
the condition of a houseplant or outdoor container plant which has outgrown its container, with the roots filling every niche of the container. Eventually, the roots will themselves strangle the plant by constricting the flow of nutrients. A hook or blade is required to cut through and separate the tangled matted outer roots when removed from the pot and before placing in the new container.

- **rooting hormone**
a chemical in powder or liquid form which promotes the formation of roots at the base of a cutting. Contains hormones and anti-fungus growth prohibitors.

- **rootlet**
a small, secondary root.

• **rootstock**
1. a rhizome.
2. the root system and lower portion of a woody plant to which a graft of a more desirable plant is attached.

• **rootworm**
the larvae of a beetle of the genus Diabrotica that feed on the roots of various plants, particularly corn.

• **roridulous**
dewy, covered with waxy plates which appear to be dew drops.

• **rosarian**
those hobbyists or professionals who specialise in the cultivation of roses.

• **rose**
apart from the flower, the term refers to the nozzle with fine holes put on the spout of a watering can.

• **rose dethorner**
florist tool to remove the thorns from roses and other flowers.

• **rose hip**
a smooth, rounded, fruit-like structure consisting of the cup-like calyx enclosing seed-like fruits.

• **rosette**
a cluster of leaves which grows in a circular overlapping pattern somewhat like the petals of a rose arising basally from a crown, as on a dandelion, or apically from an axis, as on many palms.

• **rostellate**
having a short beak.

• **rostellum**
a slender growth of tissue located at the upper part of the column which physically seperates the male and female parts thus providing a barrier to prevent self pollenisation. The rostellum also is used to apply sticky 'glue' to the back of the pollinator (usually an insect such as a bee) to attach the Pollinarium (the pollen transport system).

• **rostrate**
1. having a beak.
2. describes an operculum with a long beak.

• **rosulate**
in the form of a rosette.

• **rot**
the softening, discolouration, and often disintegra-

tion of a succulent plant tissue as a result of fungal or bacterial infection.

- **rotary edger and trimmer**
  a lawn tool which has rotating spikes on one edge to trim grass back from drives and walks.

- **rotary mower**
  a mower with a blade that spins in a horizontal plane from a central rod. Its advantages are the ability to cut tall grass, versatility of movement, a less expensive purchase price and blades that can be easily sharpened.

- **rotate**
  describes a corolla that is wheel-shaped flat and circular in outline.

- **rotation**
  specifically crop rotation: changing the plants in the same growing area. This will decrease the soil borne diseases.

- **rotation grazing**
  alternating between two or more pastures.

- **rotenone**
  material used a lot by organic gardeners. It is derived from the roots of tropical legumes. It does break down in sunlight and the side effect is that it is toxic to good and bad insects.

- **rotund**
  rounded

- **rough**
  papillose.

- **roundworm**
  nematode

- **row cover fabric**
  a loosely woven translucent fabric used to keep insect pests off crops. It also functions as a cloche.

- **row covers**
  any type of semitransparent materials used to cover plants, trap heat, enhance growth, and provide protection from frost or winds. Commercial growers sometimes use this method.

- **row sprigging**
  planting of sprigs in rows or furrows.

- **r-strategy**
  ecological strategy where organisms rely on high reproductive rates for continued survival within the community. Populations of r-

strategists are subject to extreme fluctuations.

- **rubefacients**
  herbal medicines which open the small blood vessels of the skin to increase its heat and redness.
- **ruderal**
  growing in poor land or waste places.
- **rudimentary**
  partially developed.
- **rufescent**
  reddish-brown.
- **rugose**
  wrinkled, leaf veins deeply etched into the upper surface of the leaf.
- **ruminate**
  having a surface which is coarsely wrinkled, appearing as though chewed by a cow and then spit out, but not with the margins appearing gnawed as in erose margins
- **run**
  a plant that runs will be growing rapidly underground. Great for ground cover but some plants take over an area by running.
- **runcinate**
  sharply incised, with the segments directed backward.
- **runner**
  a thin stem that grows horizontally above ground and it roots and grows new plants at the nodes, e.g. strawberries. Not to be confused with stolon.
- **run-off**
  when liquids (such as in watering an area of ground or a fast rain) washes off quickly a run-off is created. Often pesticides and fertilisers are washed into waterways from lawn and garden run-off.
- **rupicolous**
  dwelling in or on rocks or stones.
- **rush**
  any of various monocotyledonous, grass-like, often tufted marsh plants of the family Juncaceae, having cylindrical often hollow stems.
- **russet**
  brownish, roughened areas on the skin of fruit as a result of cork formation.

- **rust**
  1. fungus of the class Urediniomycetes.
  2. any of various plant diseases caused by members of the Urediniomycetes or by species of Albugo. The diseases are called 'rusts' because many of the causal agents form rust-coloured spores on affected plants.

- **sac**
  a cavity or chamber, locule.

- **saccate**
  with a conspicuous hollow swelling. The term is usually used to describe the bag, pouch, or sac-shape of the lip on an orchid flower, like the lip-shape of species in genus Paphiopedilum.

- **saccharomyces**
  reported to be allergenic. Baker's Yeast.

- **sagittate**
  shaped like an arrow-head, the basal lobes directed downward.

- **saline soil**
  soil containing sufficient soluble salt to adversely affect the growth of most crop plants.

- **salinity**
  refers to an excess of salts in the soil which causes leaves to scorch and turn yellow and does great harm to many plants.

- **salt marsh hay**
  probably a better mulch that is collected from the grasses grown in coastal marshes. It generally does not contain weed seeds and is fluffy.

- **saltation**
  a mutation occurring in the asexual state of fungal growth, especially one occurring in culture.

- **salverform**
  having a slender tube which expands abruptly.

- **salver-shaped**
  describes a corolla that has a slender tube abruptly expanded into a flat limb.

- **samara**
  a dry, one-seeded, indehiscent winged fruit, a key.

- **samaroid**
  resembles a samara, a winged fruit.

- **sand**
  soil particle between 0.05 and 2.0 mm in diameter.

- **sandbar**
  a temporary ridge of sand formed by currents in a body of water.

- **sanitation**
  principle of plant disease control involving removal and burning of infected plant parts and decontamination of tools, equipment, hands, etc.

- **sanitisation**
  elimination of pathogenic or deleterious organisms, insect larvae, intestinal parasites, and weed seeds.

- **sap**
  the fluid in plants. Most known is the collecting of maple sap made into syrup.

- **saphrophyte**
  plants often lacking chlorophyll, receiving nourishment from dead or decaying organic matter, needing the services of certain fungi to be able to absorb food.

- **sapling**
  a very young tree.

- **saponin**
  a glycoside compound of plants which has a foaming or soap-like action.

- **saprobe**
  an organism that obtains its nutrients from non-living organic matter (commonly dead and decaying plant) or animal matter by absorbing soluble organic compounds. Also known as Saprotroph.

- **saprogenesis**
  survival, that phase of the life cycle of a pathogen during which it is not actively causing disease in a living suscept.

- **sapropel**
  a gaseous product of decomposition of organic depositions on the bottom of a body of water, especially common in marshes, formed under anaerobic conditions, and with the fetid odour of rotten eggs.

• **saprophyte**
non-parasitic nutritional mechanism by which an organism obtains its food exclusively from the degradation of non-living organic material.

• **saprophytic**
feeding from dissolved or decayed organic material, as a saprophyte.

• **saprotroph**
see **saprobe**.

• **sapwood**
the more recently produced, outer layer of wood tissue of a tree, usually lighter in colour than the heartwood.

• **sarcoay**
a hyperplastic symptom in which swellings occur above and below portions of organs that are tightly encircled.

• **sarmentose**
with long slender stolons.

• **satellite virus**
a defective virus requiring a helper virus to provide functions necessary for replication. It may code for its own coat protein or various other products.

• **saturation cratering**
ancient land around a volcanic area, filled with old and new craters from repeated eruptions.

• **savanna**
grassland with scattered trees or shrubs, often a transitional type between true grassland and forestland, and accompanied by a climate with alternating wet and dry seasons.

- **sawflies**
larvae of various insects that feed on leaves, sometimes rolling them.

- **saxatile**
growing among or upon rocks.

- **saxicolous**
dwelling in or near rocky places, growing on rocks

- **scab**
any of a wide range of unrelated plant diseases having a roughened, crust like diseased area on the surface of a plant organ (e.g., apple scab, potato scab, wheat scab).

- **scaberulous**
slightly scabrous.

- **scabrid**
minutely rough.

- **scabridulous**
slightly rough.

- **scabrous**
rough with short, rigid projections.

- **scaffold branches**
the main side branches of a tree, especially a fruit tree.

- **scalariform**
having a ladder-like pattern.

- **scald**
when plants have an overexposure to sunlight a discolouration will develop. Often when plants do not become acclimated slowly into a sunny location from being indoors or from the shade.

- **scale**
1. a thin, membrane-like covering of the bud or twig base. A fine grain-like surface material.
2. sap-feeding insects that live under shell-like scales on stems, bark, and leaves.

- **scale leaves**
specialised leaves, including those covering buds or composing bulbs.

- **scalping**
1. in seed processing: removal of material larger than the crop seed.
2. of turf: removal of an excessive quantity of green shoots at any one mowing.

resulting in a stubbly, brown appearance due to the exposed stems, stolons, and dead leaves.

- **scaly**
  having tiny scales attached at only one end.

- **scandent**
  climbing, but not self-supporting.

- **scape**
  a leafless main flower-stalk arising from the underground or sub-surface parts of a plant (species of Paphiopedilum are a good example). It may bear scales, bracts, and may be one or many-flowered

- **scapose**
  bearing or resembling a scape.

- **scarification**
  the physical or chemical treatment given to some seeds in order to weaken the seed coat sufficiently for germination to occur.

- **scarify**
  to scratch or break the hard coat of some seeds, so they will germinate easily. Fine sandpaper or just soaking the seed will do the trick.

- **scariose**
  scarious, having thin, dry, shrivelled tissues.

- **scarred**
  having marks where organs have fallen off, such as leaf scars on stems.

- **schizocarp**
  a fruit that splits between carpels into one-seeded portions or lobes.

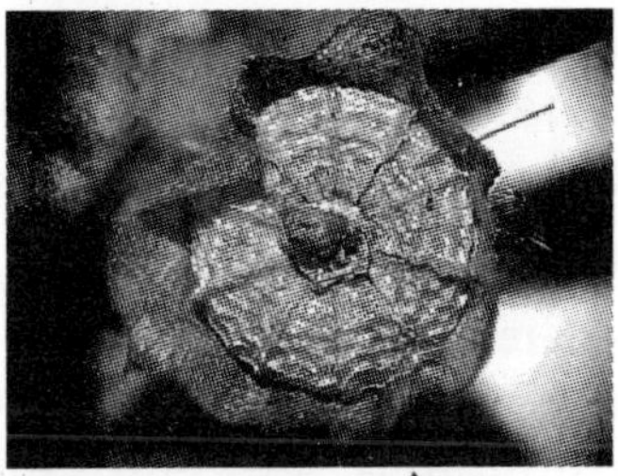

- **scientific name**
  the internationally recognised Latin name of a plant that will be descriptive of the feature of the plant, or will commemorate a person connected with it. The name of the species will consist of two parts, the genus name and the species name. This system was first started in 1753 by a Swedish botanist, Linnaeus.

- **scion**
  a cutting from the upper portion of a plant that is

grafted onto the rootstock of another plant, usually a related species.

- **scion rooting**
covering a low graft with soil so that the plant develops roots directly from both the rootstock and the scion.

- **scion wood**
shoots from which graft scions are cut.

- **sclereid**
a variably shaped sclerenchymatous cell of a higher plant.

- **sclerenchyma**
tissue composed of cells with thickened and hardened walls.

- **scleromorph**
a plant with leaves and/or stems that are hard, usually having a thick cuticle.

- **sclerophyll**
a woody plant with small, leathery, evergreen leaves and generally found in hot, dry climates

- **sclerophyllous**
having leaves stiffened by sclerenchyma.

- **sclerotium**
hard, resistant, multicellular resting body, usually with a differentiated cortex and medulla, that under favourable conditions can germinate to produce mycelium or sexual or asexual fruiting bodies. (pl. sclerotia)

- **sclerotized**
in nematodes: hardened refractive regions.

- **scolecospore**
a very long, thin spore (with a length/width ratio more than 15:1).

- **scolecosporous**
having long, thin (filiform spores).

- **scoop shovel**
a shovel with a very large blade, often made of thin light-weight metal or plastic, used for moving large amounts of crops such as ear corn, or clearing water.

- **scooping**
  a technique that is used to produce a shorter and bushier plant by scraping out the centre eye of the bulb.

- **scopulariopsis**
  it may produce arsine gas if growing on arsenic substrate.. This can occur on wallpapers covered with Paris green. It has been found growing on a wide variety of materials including house dust. It is associated with type III allergy.

- **scorch**
  injury to leaves due to lack of sufficient water, excessive transpiration or injury to the water-conducting system of the plant.

- **scorching**
  see **scald**.

- **scorpioid**
  describes an inflorescence that is circinately coiled while in bud.

- **scrambler**
  a plant with long stems, often with thorns to help it support itself as it climbs through shrubs.

- **scree**
  fragmented rocks usually found at the base of a cliff or boulder where the action of wind and water has broken the particles loose. Scree may be used in gardens as mulch.

- **screen**
  a single plant or grouping of plants used to bar certain parts of the landscape from view.

- **scuffle hoe**
  a medium-weight hoe with the blade fastened to the handle in the middle of the horizontal blade, allowing it to be pushed and pulled, cutting weeds on each direction.

- **scurfy**
  covered with tiny, broad scales.

- **scutate**
  platter-shaped, with a flat centre and raised margins.

as a water lily leaf, *Nymphaea lotus*.

- **scutelliform**
shaped like a small shield.
- **scutellum**
in nematodes: an enlarged, shield-like plasmid.
- **scythe**
a long-handled blade for making grass into hay, the term is sometimes used for the blade without the handle.
- **sealed storage**
storage in a sealed container, usually refers to airtight (hermetic) storage.
- **seawater**
ocean water with a high salt content.
- **secateur**
hand-held pruning shears.
- **secondary cycle**
of plant disease: any cycle initiated by innoculums generated during the same season
- **secondary infection**
any infection caused by innoculums produced as a result of a primary or a subsequent infection, an infection caused by secondary innoculums.
- **secondary innoculums**
innoculums produced by infections that took place during the same growing season.
- **secondary metabolite**
product of intermediary metabolism released from a cell, such as an antibiotic.
- **secondary organism**
an organism that multiplies in already diseased tissue but is not the primary pathogen.
- **secondary rot**
rot caused by a secondary organism.
- **secondary species**
the species subordinate to the dominant species, like dogwood, Cornus, in an oak, Quercus, forest.
- **secondary symptom**
a symptom of virus infection appearing after the first (primary symptoms).
- **secondary tillage**
tillage that works the soil to a shallower depth than primary tillage, providing additional pulverisation, levels

and firms the soil, closes air pockets, kills weeds.

- **secund**
  to one side, as flowers on an inflorescence.

- **sedatives**
  medicines which can slow certain bodily functions and promote sleep.

- **sedentary**
  staying in one place, stationary.

- **sedge**
  any of the grass-like, monocotyledonous plants of the family Cyperaceae, having achenes and solid stems which are triangular in cross-section.

- **seed**
  1. a mature (ripened) ovule consisting of an embryonic plant and a store of food (stored in the endosperm, in some species), all surrounded by a protective seed coat.
  2. to sow or plant seed (e.g., broadcasting or drilling of small-seeded grasses and legumes or other crops).

- **seed coat**
  the outer protective covering of a seed.

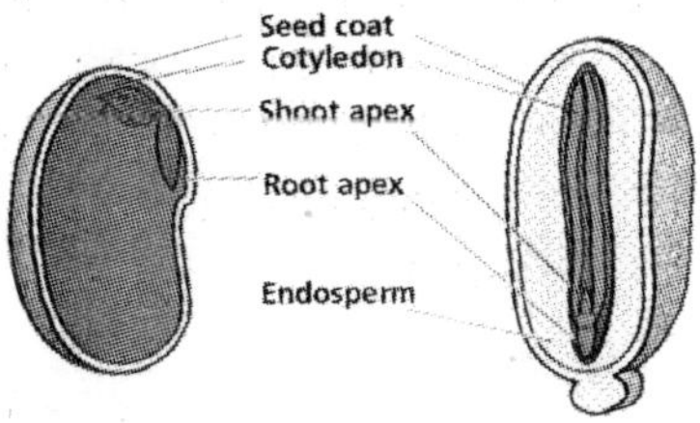

- **seed cotton**
  cotton, as harvested and before ginning, consisting of seeds with the fibres attached and usually including measurable amounts of foreign matter.

- **seed drill**
  1. a shallow trench in which to sow seeds.
  2. an implement for planting seeds.

- **seed fern**
  extinct fern-like plants that bore seeds on the frond.

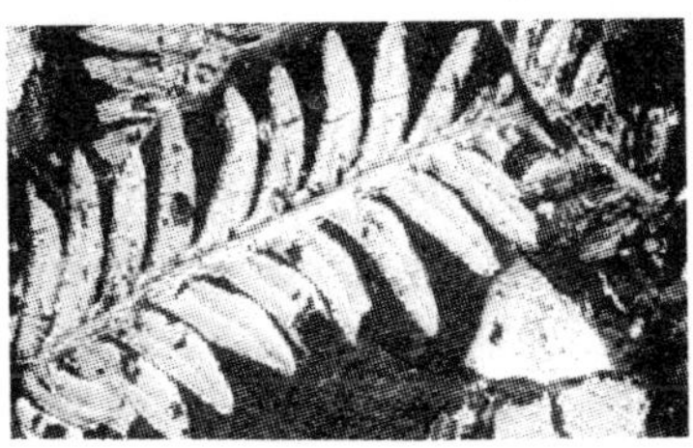

- **seed index**
  the weight in grams of 100 seed.
- **seed leaf**
  a cotyledon.
- **seed lot**
  seeds of a particular crop gathered at one time and likely to have similar germination rates and other characteristics.
- **seed mat**
  a fabricated mat with seed (and fertiliser) applied to one side, the mat serves as a vehicle to apply seed and fertiliser, control erosion, and provide a favourable microenvironment for seed germination and establishment.
- **seed multiplication**
  the aggregate of all practices required to grow a plant to maturity and produce seed, including those practices necessary for harvesting and preparing seeds for subsequent plantings.
- **seed processing**
  the operations involved in preparing harvested seed for marketing.
- **seed set**
  the formation of seeds or seed pods on the parent plant.
- **seed stocks**
  see **stock seed**.
- **seed vigour**
  those seed properties that determine the potential for rapid uniform emergence and development of normal seedlings under both favourable and stress conditions.
- **seed bank**
  a facility designed for the in situ conservation of individual species and varieties through controlled seed storage.
- **seedbed**
  an area cultivated for planting seeds.
- **seedhead**
  the dry fruit containing ripe seeds.

- **seedling**
  a young plant grown from a seed.

- **seedling, normal**
  a seedling possessing the essential structures indicative of the ability to produce plants under favourable conditions.

- **segment**
  one of the parts of a leaf or other like organ that is cleft or divided.

- **segregate**
  1. a new genus culled from an established one for taxonomic reasons.
  2. a hybrid which exhibits the genetics of one of the parents more clearly than those of the other parent.

- **seiche**
  a rapid and sometimes violent fluctuation in water level within a landlocked body of water, usually due to sudden changes in barometric pressure, the ripples lasting from minutes to several hours.

- **selected seed**
  selected seed is a class of tree seed that is the progeny of rigidly selected trees or stands of untested parentage that have promise but not proof of genetic superiority, and for which geographic source and elevation shall be stated on the certification label.

- **selection**
  a distinct form of a plant, sexual or asexual in origin, selected and propagated for monetary reasons or beauty, and named as a cultivar.

- **selective felling**
  when only choice trees are cut in a forest.

- **selective herbicide**
  a pesticide that kills only one type of plant, for example broadleaf herbicides only kill broadleaf weeds, not turf grasses.

- **selective medium**
  medium that allows the growth of certain types of micro-organisms in preference to others. For example, an antibiotic-containing medium allows the growth of only those micro-organisms resistant to the antibiotic.

- **self-pollination**
  the transfer of pollen from one flower to another flower on the same plant.

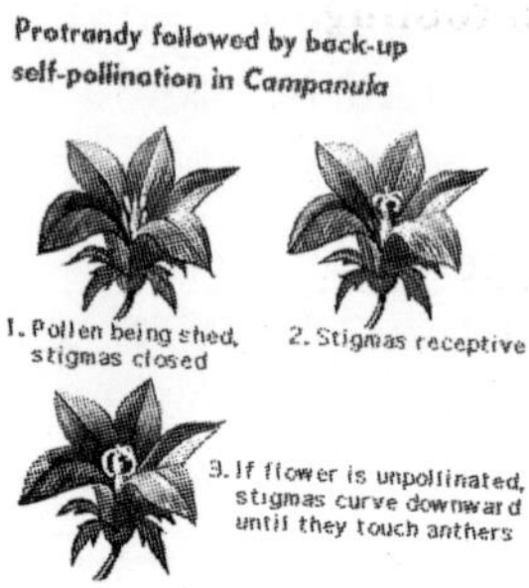

- **self-seed**
  plants releasing seeds spontaneously which grow into new plants.

- **self-cleaning**
  herbaceous plants that drop spent blossoms, thus not requiring deadheading.

- **self-clinging**
  refers to plants which climb rocks or structures without the support of other plants, perhaps by suckers or tendrils.

- **self-colour**
  a flower with single coloured petals.

- **self-fertile**
  able to be pollinated by its own flowers.

- **selfing**
  the pollination by the plant's anthers of either the same flower or a flower on the same plant. In hybrid names, you will often see (x self) in the name of the plant which means it was crossed by the same plant.

- **self-pollination**
  the transfer of pollen from the anther of a flower to the stigma of the same flower, or to different flowers on the same plant.

- **self-seeded**
  a plant's habit of shedding seeds in the immediate area. They will then germinate without outside help. Many annuals use self-seeding (i.e. cleome-marigolds).

- **self-sowing**
  produces offspring from seed without intervention.

- **semelparity**
  refers to an organism that has a single reproductive season and then dies.

- **semiarid turfgrass**
  turf grasses adapted to grow and persist in semiarid regions without irrigation

(e.g., buffalo grass, gram grasses, and wheat grasses).

- **semi-double**
having more than the usual number of petals but with at least some pollen-producing stamens.

- **semi-evergreen**
retaining at least some green foliage well into winter, or shedding leaves only in cold climates.

- **semi-hardwood cutting**
a cutting taken from a stem that has just begun to develop woody tissue, for the purpose of propagation.

- **semi-hardy**
questionably hardy in a given temperature zone, susceptible to damage by extreme cold or heat.

- **semi-ripe cuttings**
cuttings taken in mid-summer to autumn to propagate woody plants.

- **semi-rotate**
saucer-shaped, almost flat.

- **semi-reiterate**
refers to a cylinder which is rounded on one side, but flat on the other; as some leaf petioles.

- **senescence**
1. the developmental stage during which deterioration occurs leading to the end of functional life of an organism or organ. Sometimes defined from specific criteria such as a decline in chlorophyll or dry weight.
2. more generally, a slowing in the rate of growth of a plant or plant organ, usually due to old age.

- **sensu**
meaning 'in the sense of' in taxonomy.

- **sensu lato**
in a broad sense

- **sensu stricto**
in a narrow sense

- **sepal**
a division of a calyx, one of the outermost circles of

modified leaves surrounding the reproductive organs

Parts of a Flower

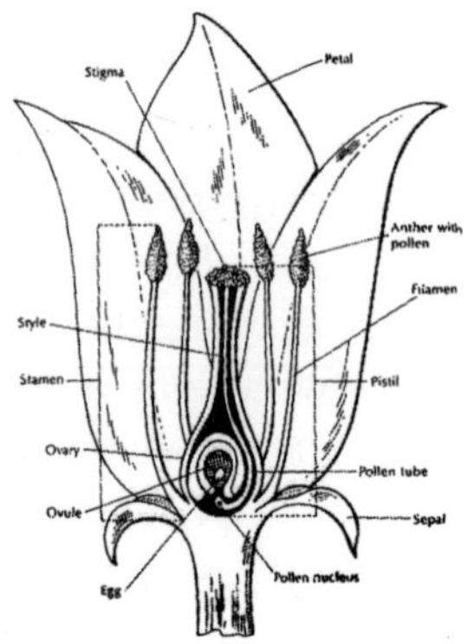

- **sepaline**
  refers to sepals.
- **sepaloid**
  like a sepal.
- **sept**
  a small dent or cavity.
- **septate**
  divided by partitions.
- **septicidal**
  describes a capsule that is dehiscing through the partitions and between the cells.
- **septifragal**
  describes the dehiscence of a fruit where the valves or backs of the carpals break away leaving the septa intact.
- **septum**
  cross wall (partition) dividing a parent cell into two daughter cells during binary fission or occurring between adjacent cells in hyphae.
- **sequence**
  the order of nucleotides in RNA or DNA or of amino acids in a polypeptide.
- **sequencing**
  determining a sequence of a nucleic acid or protein.
- **sere**
  the complete cycle of changes from the original state to the climax condition, e.g., oak-maple forest, forest destroyed by fire, meadow, brush land, young mixed conifer-deciduous forest, oak-maple forest.
- **serial dilution**
  series of stepwise dilutions (usually in sterile water) performed to reduce the populations of micro-organisms in a sample to manageable numbers.
- **seriate**
  in a whorl or row, generally preceded by a number or numerical prefix, as

triseriate meaning in three rows.

- **sericeous**
bearing fine, usually straight, apprised hairs.

- **series**
a group of similar, but not identical, plants.

- **serology**
branch of science dealing with properties and reactions of sera, particularly the use of antibodies in the sera to examine the properties of antigens.

- **serotinal**
appearing in the autumn or late summer.

- **serotype**
a subdivision of virus strains distinguished by protein or a protein component that determines its antigenic specificity

- **serpula lachrymals**
common cause of extrinsic asthma (immediate-type hypersensitivity: type I). Acute symptoms include oedema and bronchi spasms, chronic cases may develop pulmonary emphysema.

- **serrate**
having sharp teeth pointing forward.

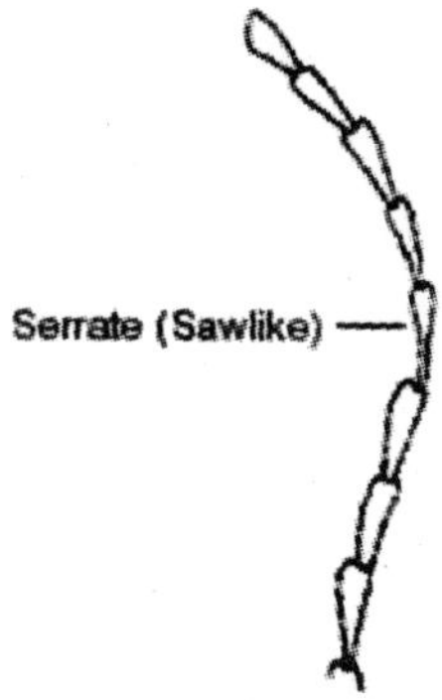

- **serrulate**
finely serrate.

- **serum**
the fluid fraction of coagulated (clotted) blood.

- **sessile**
a stalk less leaf or flower which is borne directly on the stem.

- **set**
1. the development of fruit, and/or seed, following pollination, to set seed.
2. to transplant as seedlings.

3. to apply as a graft.
4. a young bulb, tuber, or other type of vegetative propagule ready for planting.
5. transformation of cement paste or concrete from a fluid-like consistency to a stiff mass.

- **set seed**
to produce seeds after flowering. In some cases a gardener wishes to prevent this in an effort to prolong flowering, or continue harvesting the foliage of an herb.

- **set stocking**
the practice of allowing a fixed number of animals on a fixed area of land during the time when grazing is allowed.

- **seta**
1. in nematodes: elongated cuticular structures articulating with the cuticle. In general, tactile sensory organs usually located around the oral openings.
2. In fungi: a bristle-like structure in some types of fruiting bodies (pl. setae).

- **setaceous**
bristle-like.

- **setiform**
bristle-shaped.

- **setose**
bristly, beset with bristles.

- **setulose**
having minute bristles.

- **sewage sludge**
the solid matter that settles out during the treatment of sewage.

- **sextant**
an instrument used mostly at sea, where there are no bench marks, to determine location by latitude and longitude.

- **sexual dimorphism**
a pronounced difference in the morphologies of the two sexes within a species.

- **sexual reproduction**
production of new generations involving the exchange of chromosomes from both a male and female parent.

- **shade**
light diminished by some blocking feature.

- **shade tree**
a larger tree planted in a location where the canopy

provides shade at ground level.

- **sharp-pointed**
acute, mucronate, pungent.

- **shattering**
1. a condition whereby seeds drop to the ground prior to harvest.
2. a method of turf cultivation involving fragmentation of a rigid or brittle soil mass, usually by a vibrating mechanical mole device.

- **shear**
a method of pruning in the landscape. Often in reference to hedges.

- **sheath**
the tubular base of the leaf surrounding the flower spike.

- **sheathing**
enclosing as by a sheath.

- **sheet composting**
a method of piling un-decomposed organic materials over the soil and waiting for decomposition. At times working it in the soil.

- **sheet mulch**
using plastic sheeting to cover ground to retain moisture and suppress weeds - see **mulch**.

- **shock symptoms**
the severe, often necrotic symptoms produced on the first new growth following infection with some viruses. Also called Acute Symptoms.

- **shoot**
a young, actively growing twig or stem.

- **short day plant**
a plant which requires light for a shorter period than it would normally receive from daylight in order to induce flowering, e.g., Chrysanthemum and Poinsettia.

- **short-day**
of, relating to, or being a plant which flowers only after being exposed to daylight for periods shorter

than a particular length, usually in the fall or spring.

- **short-lived**
  describes a plant, usually perennial, where the individuals live for only a few seasons.
- **short-lived perennial**
  turf grasses normally expected to live only 2 to 4 years.
- **shot hole**
  a symptom in which small, diseased fragments of leaves fall off, leaving small holes in their place.
- **shovel**
  1. a long-handled scoop-like implement used for digging, lifting, and throwing, etc.
  2. the quantity contained in one scoop.

- **shreddy**
  dividing into fragile, thin, narrow sheets.
- **shrub**
  a woody plant that doesn't have a single trunk and whose branches originate from or near the base.

- **shrubland**
  land on which the vegetation is dominated by shrubs.
- **shrublet**
  a dwarf woody plant.
- **sialagogues**
  herbal medicines that can increase the flow of saliva.
- **sibbed**
  plants that have the same parentage. In hybrid names, you may see (x sib) in the name. This means the cross of the plant was made using the same parents.
- **sickle**
  a short-handled hand tool similar to a scythe but

smaller and curved into a semicircle, used to cut heavier crops like corn.

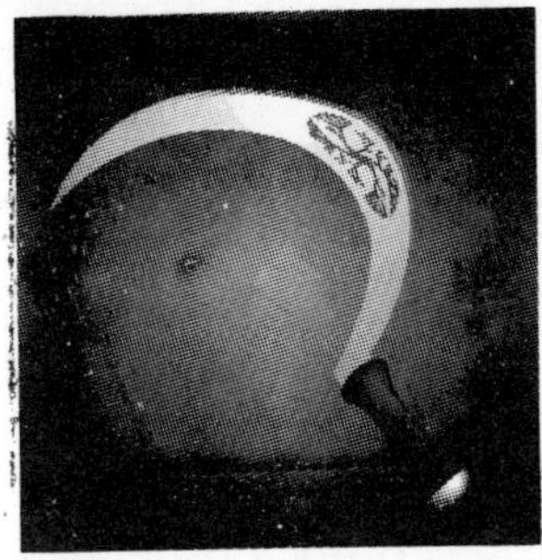

- **sickle bar**
  a mechanical cutting tool that fastens onto a tractor's power source.

- **side dress**
  when one fertilises above the ground without ploughing it. Placing the fertiliser on the side of the plant material. Sometime side dressing is used to encourage growth during the season.

- **siderophore**
  non-porphyrin metabolite secreted by certain micro-organisms that forms a highly stable coordination compound (chelate) with iron, a high-affinity iron-binding compound. There are two major types: catecholates and hydroxamates.

- **sieve**
  a garden sieve is a frame with a mesh bottom. Mainly used for separating compost, but sometimes-used in very stony gardens. Home made ones made out of two by fours, 2' x 2' with a $^1/_2$ inch hard cloth bottom is most handy.

- **sieve plate**
  perforated wall area between two phloem sieve cells through which they are connected.

- **sieve tube**
  a series of phloem cells forming a long cellular tube through which food materials are transported.

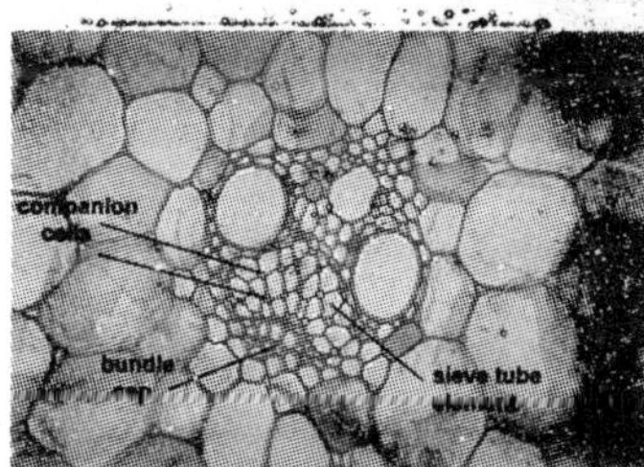

- **sigmoid**
  doubly curved in opposite directions, like the Greek letter sigma.

- **sign**
  a visible manifestation of a causal agent of plant dis-

ease (e.g., fungal spores or other fungal structures, bacterial ooze).

- **silage**
  forage preserved in a succulent condition by partial anaerobic, acid fermentation.

- **silage, additive**
  material added to forage at the time of ensiling to enhance favourable fermentation process.

- **siliceous**
  containing silica.

- **silicle**
  a short silique, no more than twice as long as it is broad.

- **silique**
  a dry, dehiscent, elongated fruit formed from a superior ovary of two carpals, with two parietal placentas and

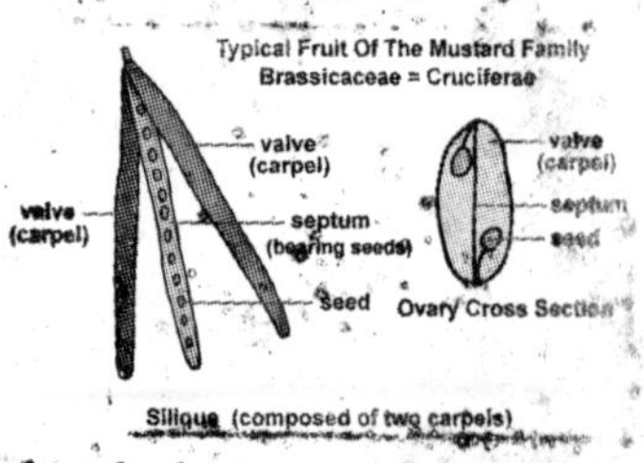

divided into two loculi by a false septum between the placentas, occurring in plants of the family Cruciferae.

- **silky**
  covered with close-pressed soft and straight pubescence.

- **silt**
  fine-grained sediment usually 1/20 millimetre or less in diameter, finer than sand particles and coarser then those composing clay.

- **silt load**
  the particles finer than sand which are carried in a suspended state in moving water.

- **silvicolous**
  growing in woodlands.

- **silviculture**
  the cultivation and care of forest trees.

- **simple**
  of one piece, not compound.

- **simple fruit**
  a fruit developed from a single ovary.

- **simple leaf**
  a leaf with an undivided blade.

- **single**
  1. a single flower.

2. describes varieties having only one bloom per stem.

• **single cross**
the first generation of a cross of two inbred lines, an inbred line and a foundation backcross, or of two foundation backcrosses.

• **single flower**
a plant containing the normal number of petals. Double flowers often have multiples of the normal number: 6 or 9 for three-petal flowers, 8 or 12 for four-petals flowers, etc.

• **single-seed descent**
a pedigree method for producing inbred lines in which each resulting line can be traced through a single plant in each preceding generation of inbreeding.

• **single-worked**
grafted only once so that the grafted plant consists of rootstock and scion.

• **sinuate**
with the outline of the margin strongly wavy.

• **sinuous**
wavy like the path of a snake.

• **sinus**
the cleft or recess between two lobes.

• **site**
1. in ecology, area described or defined by its biotic, climatic, and soil conditions as related to its capacity to produce vegetation.
2. area sufficiently uniform in biotic, climatic, and soil conditions to produce a particular climax vegetation.

• **site-directed mutagenesis**
insertion of a different nucleotide at a specific site in a molecule using recombinant DNA methodology.

• **skeletonised**
refers to a leaf which has only ribs and veins left, the soft tissue having been consumed by insects or bacteria.

• **skylight**
the diffuse and reflected light of the sky.

• **slice seed**
a technique used to sow seed. A machine cuts or slices grooves into the lawn or soil and drops seeds directly into the grooves. It is used to fill in a thinning

lawn without disturbing the existing grass excessively.

- **slicing**
  a method of turf cultivation in which vertically rotating, flat tines slice intermittently through the turf and the soil.

- **slime layer**
  diffuse layer of polysaccharide exterior to the cell wall in some bacteria..

- **slime mould**
  non-phototropic eukaryotic micro-organism lacking cell walls, which aggregate to form fruiting structures (cellular slime moulds) or simply masses of protoplasm (a cellular slime mould).

- **slip**
  an older fern once used for a cutting. Slips are taken for propagation.

- **sliver**
  a continuous strand of loosely assembled fibres, approximately uniform in cross-sectional area and without twist.

- **sliver, drawing**
  a sliver produced by a draw frame.

- **slow-release fertiliser**
  fertiliser with a rate of dissolution less than is obtained for completely water-soluble fertilisers, may involve compounds that dissolve slowly, materials that must be decomposed by microbial activity, or soluble compounds coated with substances highly impermeable to water. Synonyms: controlled-availability, controlled-release, delayed-release, metered-release, and slow-acting fertiliser.

- **sludge**
  precipitate of solids from water or sewage treatment, which may or may not contain harmful compounds, generally composted for a year or two before applying as organic fertiliser.

- **slump test**
  a test used to determine workability of cement or concrete.

- **small fruits**
  those species which produce edible fruits, but do not grow on trees, e.g., strawberries and blueberries.

- **smoke jumper**
  a trained fire-fighter who is dropped from an airplane in the path of a wildfire to attempt to control the blaze.

- **smooth**
  without roughness or pubescence.

- **smut**
  any of a number of plant diseases caused by the smut fungi (Ustilaginales), characterised by masses of dark, powdery, and sometimes odorous spores (e.g., stinking smut of wheat, common smut of maize).

- **smut spore**
  a dark, thick-walled resting spore of a smut fungus, may germinate to produce a promycelium, often improperly termed a chlamydospore.

- **snag**
  1. part of the rootstock left during a graft to hold the scion in place.
  2. removal of the projection when the scion no longer needs it.
  3. a rough projection on a woody plant which needs to be removed.

- **snath**
  the handle of a scythe.

- **snow fencing**
  wooden or plastic lath joined by wire into long rolls, then erected 100 feet (33 m.) or so on the windward side of a road or barnyard. The lath breaks the velocity of the wind allowing snow particles to drop before reaching the road. It may also be used as a shade cover for plants.

- **soaker hose**
  hoses that have hundreds of mini holes to let the water out slowly and can be left on for a long period of time. Great for vegetable gardens and beds that need to be watered frequently.

- **sobole**
  a shoot, especially from the ground.

- **soboliferous**
  bearing soboles.

- **social dominance**
  the control of behaviour of those animals lower in the hierarchy, by aggression, intelligence, or other factors of the alpha animals.

- **social facilitation**
  the effect of the closeness of one organism upon the conduct of another one.

- **sod coring tool**
  a lawn tool which removes spikes of soil, about one half inch (1 cm) in diameter, to allow air and water to penetrate.

- **sod cutter**
  a device to sever turf from the ground, the length and thickness of the cut are adjustable.

- **sod heating**
  heat accumulation in tightly stacked sod, may reach lethal temperatures.

- **sod planter**
  a lawn tool which cuts a hole of about 2 inches (5 cm.), releases the core with a hand or foot system, picks up a new sod block and inserts it.

- **sod production**
  culture of turf to a quality and maturity that allows harvesting and transplanting. Synonym: sod rooting.

- **sod strength**
  the relative ability of sod to resist tearing during harvesting, handling, and transplanting.

- **sodding**
  planting turf by laying sod.

- **sodic**
  of, relating to, or containing sodium.

- **softwood**
  the immature stems of woody plants. Also, term applied to conifers as opposed to the broad-leaved trees, or hardwoods.

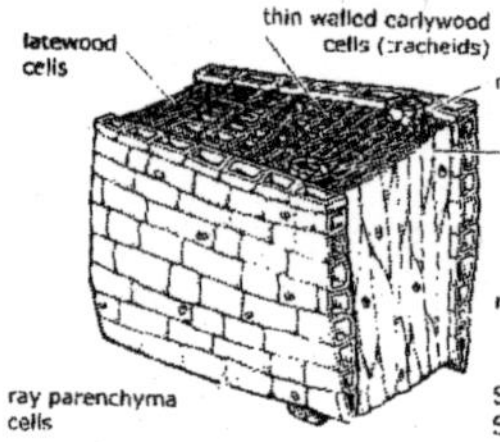

SECTION OF SOFTWOOD

- **softwood cutting**
  a non-woody piece of a woody plant that is cut from the stock plant to asexually propagate a new individual plant.

- **soil**
  1. unconsolidated mineral or material on the immediate surface of the earth that serves as a natural medium for the growth of land plants.

2. unconsolidated mineral or organic matter on the surface of the earth that has been subjected to and influenced by genetic and environmental factors of: parent material, climate (including water and temperature effects), macro -organisms and micro -organisms, and topography, all acting over a period of time and producing a product—soil—that differs from the material from which it is derived in many physical, chemical, biological, and morphological properties and characteristics.

- **soil aggregate**
  unit of soil structure generally less than 10 mm in diameter and formed by natural forces and substances derived from root exudates and microbial products which cement smaller particles into larger units.

- **soil amendment**
  anything added to the soil to improve the present situation, i.e. drainage, nutrients, or makeup.

- **soil atmosphere**
  gases occupying the pore space in soil. Generally characterised as having a greater percentage of carbon dioxide and a lesser percentage of oxygen than the overlying air.

- **soil biochemistry**
  branch of soil science concerned with enzymes and the reactions, activities, and products of soil micro-organisms.

- **soil conditioner**
  any material added to soil to improve its structure, texture, tilt or drainage.

- **soil cuber**
  hand-held device into which moist soil is packed, then released in the form of a cube or block in which a single seed is planted.

- **soil extract**
  solution separated from a soil suspension or from a soil by filtration, centrifugation, suction or pressure.

- **soil fixation**
  the conversion of a soluble substance such as phosphorus from the exchangeable form useful to plants, to a relatively insoluble form.

- **soil horizon**
  layer of soil or soil material approximately parallel to the land surface and differing from adjacent genetically related layers in physical, chemical, and biological properties or characteristics such as colour, structure, texture, consistency, kinds and number of organisms present, and degree of acidity or alkalinity.

- **soil less gardening**
  another name for hydroponics. Gardening in something other than soil or water and rocks.

- **soil less mix**
  this would be any medium for containers. The substances would be like peat moss, prelate, vermiculite, or materials like sand or bark. However, fertilising is a must since none of these have any real nutrient quality.

- **soil microbiology**
  branch of soil science concerned with soil-inhabiting micro-organisms and their functions and activities.

- **soil mix**
  a prepared mixture used as a growth medium for turf grass.

- **SOM**
  Soil Organic Matter. Organic fraction of the soil exclusive of not decayed plant and animal residues.

- **soil polymers**
  super absorbent polymers that have been around for a while, that can be added to retain moisture.

- **soil population**
  1. all the organisms living in the soil, including plants and animals.
  2. members of the same taxon.

- **soil pore**
  that part of the bulk volume of soil not occupied by soil particles. Soil pores have also been referred to as interstices or voids.

- **soil probe**
  a soil-sampling tool, usually having a hollow cylinder with a cutting edge at the lower end.

- **soil quality**
  continued capacity of soil to function as a vital living sys-

tem to sustain biological productivity, maintain the quality of the environment, and promote plant, animal, and human health.

- **soil salinity**
  amount of soluble salts in a soil. The conventional measure of soil salinity is the electrical conductivity of a saturation extract.

- **soil sampler**
  a hollow tube with a T handle, which is pressed into the soil to get a core of the horizon.

- **soil science**
  science dealing with soils as a natural resource on the surface of the earth including soil formation, classification and mapping, and physical, chemical, biological, and fertility properties of soils per se, and these properties in relation to their use and management.

- **soil series**
  system of soil taxonomy, a conceptualised class of soil bodies (polypedons) that have limits and ranges more restrictive than all higher taxon. The soil series serve as a major vehicle to transfer soil information and research knowledge from one soil area to another.

- **soil shredder**
  a machine that crushes or pulverises large soil aggregates and clods to facilitate uniform soil mixing and topdressing application.

- **soil solution**
  aqueous liquid phase of the soil and its solutes.

- **soil structure**
  combination or arrangement of primary soil particles into secondary particles, units, or pads. The secondary units are characterised and classified on the basis of size, shape, and degree of distinctness into classes, types, and grades, respectively.

- **soil texture**
  relative proportions of the various soil separates in a soil. The major textural classes are sand, silt, and clay.

- **soil water potential-(total)**
  amount of work that must be done per unit quantity of pure water in order to transport reversibly and

isothermically an infinitesimal quantity of water from a pool of pure water, at a specified elevation and at atmospheric pressure, to the soil water (at the point under consideration). The total potential (of soil water) consists of the following: gravitational potential, metric potential, and osmotic potential.

- **soiling**
  see **topdressing**.
- **soilless mix**
  potting medium that contains a mixture of ingredients from the materials listed for potting medium, but no mineral soil.
- **solarisation**
  method to control pathogens and weeds where moistened soil in hot climates is covered with transparent polyethylene plastic sheets, thereby trapping incoming radiation.
- **solid**
  smoothly pithy, the twig centre neither chambered nor hollow.
- **solitary**
  borne singly or alone, not in clusters.
- **soluble fertiliser**
  a fertiliser that is mixed with water and used not only for root fertilising but can also be sprayed on the foliage. A common practice for houseplants but gardeners can use the practice in their landscape too.
- **soluble salt**
  salts from fertilisers and tap water that are dissolved in water.
- **solutio**
  Latin term for a solution.
- **solvent**
  a liquid that can dissolve a substance.
- **somaclonal variation**
  variability in clones generated from a single mother plant, leaf, etc. by tissue culture.
- **somatic**
  belonging to the vegetative part of a plant as opposed to the reproductive.
- **somatic cell hybridisation**
  production of hybrid cells by fusion of two protoplasts

with different genetic makeup.

- **sooty mildew**
  it is a fungus. It looks like dark grey spot or in general scum on stems and leaves. It is formed from the honeydew excreted by aphids and other insects. The honeydew that is created is attacked by the mildew and will go after the sugars in the honeydew.

- **sooty mould**
  a fungus of the family Capnodiaceae and of certain other families of the order Dothideales. The organisms grow epiphytically, utilising honeydew, and form dark, spongy, hyphal mats on the surfaces of certain plants.

- **soralium**
  a clump of soredia.

- **soredium**
  a microscopic clump of algal cells surrounded by the hyphae of fungi and erupting at the surface of the thallus.

- **sorus**
  a cluster or mass of spores or sporangia.

- **source-identified seed**
  source-identified is a class of true seed defined as seed from natural stands with known geographic source and elevation, or a plantation of known geographic location, as specified in the standards of the various certifying agencies.

- **southern blot**
  hybridisation of single-stranded nucleic acid (DNA or RNA) to DNA fragments immobilised on a filter.

- **southern blotting**
  procedure for transferring denatured DNA from an agarose gel to a nitrocellulose filter where it can be hybridised with a complementary nucleic acid

- **spacing**
  measured distance between the centre of one plant and the centre of the next closest ones.

- **spade**
  a digging implement, foot powered, usually of heavier construction than a shovel and often with a flat end.

• **spadix**
a fleshy flower spike in which tiny florets are embedded.

• **span length**
of cotton: The distance spanned by a specified percentage of the fibres in a fibro graph test beard.

• **spartoid**
rush-like but sharp and rigid.

• **spasmolytic**
describes an herbal preparation that can ease cramps or spasms.

• **spathaceous**
resembling or having a spathe.

• **spathe**
a large bract, sometimes highly coloured, surrounding or enclosing a spadix. The spathe flower is characteristic of the aroids, such as Anthurium and Spathiphyllum.

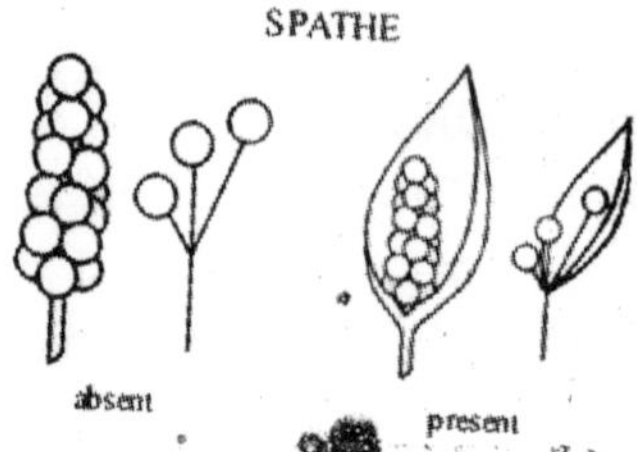

• **spatheole**
a small or secondary spathe.

• **spathiform**
shaped like a spathe.

• **spatial variability**
variation in soil properties laterally across the landscape, at a given depth, or with a given horizon, or vertically downward through the soil.

• **spatulate**
gradually narrowing downward from a rounded summit, spoon-shaped.

• **spear**
see **stylet**.

• **species**
used when naming plants. Designates a specific species of the 'Genus' and is best described as the plant words equivalent to our Christian names (or first names). Will follow the Genus name and is usually in Latin. Note:

Once a plant's full name is used, i.e. Hedera helix, future listings will abbreviate the Genus name and follow it with the species name. An example would be, H. helix, as the next plant in a listing.

- **species diversity**
  almost the same as species richness, but in more technical literature, an ecosystem is said to be more diverse if the species present have equal population sizes and less diverse if many species are rare and some are very common.

- **species epithet**
  the second word of a species name (e.g., gamines in the species Puccinia graminis)

- **species name**
  a Latin name consisting of two words, the generic name and the species epithet (e.g., Puccinia graminis)

- **species richness**
  the number of species within a region.

- **specific activity**
  amount of enzyme activity units per mass of protein. Often expressed as micromoles of product formed per unit time per milligram of protein. Also used in radiochemistry to express the radioactivity per mass of material (radioactive and non radioactive).

- **specific epithet**
  designation of a particular organism in the binomial nomenclature system. For example, coli is the specific epithet of Escherichia coli.

- **specimen**
  1. item considered typical of its class, e.g., an herbarium specimen.
  2. plant grown for landscape exhibition, e.g., a specimen tree.

- **specimen plant**
  a plant that is highlighted to show off its special qualities. Sometimes used as a focal point.

- **specimen tree**
  a tree placed so people can gain the greatest enjoyment from the colour, texture, scent, or other pleasures it provides.

- **spectroscopy**
  observation by means of an optical device (spectroscope) of the wavelength

and intensity of electromagnetic radiation (light) absorbed or emitted by various materials. Theoretical interpretation of well-defined wavelengths of elements (often present in only minute quantities) in obtained spectra leads to knowledge of atomic and molecular structure.

- **spent**
  bulbs and flowers of a plant that have finished blooming. It's time to deadhead them.

- **spermagonium**
  in certain fungi: a structure within which male reproductive cells (spermatia) are formed. In rust fungi, a spermagonium is called a pycnium.

- **spermatheca**
  in nematodes: an enlarged portion of the female gonad between the oviduct and the uterus functioning in the storage of the sperm.

- **spermatium**
  1. a non-motile male reproductive cell that can function in spermatisation
  2. the male gamete of the rust fungi.

- **spermatium**
  the non-motile male cell of red seaweeds, also used for similar cells in fungi.

- **spermatisation**
  in certain higher fungi: the union of a spermatium with a female reproductive structure

- **spermatocyte**
  in nematodes: A cell giving rise to spermatozoa or spermatozoids.

- **spermatophyta**
  a major taxonomic division containing all plants which reproduce by seed, subdivided into Gymnosperm and Angiosperm.

- **spermatozoid**
  a motile ciliated male reproductive cell.

- **spermosphere**
  area of increased microbial activity around a germinating seed.

- **sphagnum moss**
  bog mosses that are often used as soil additives and packing material because of their ability to retain moisture.

- **sphagnum moss**
  many mosses native to bogs are sphagnum. Used for the lining of hanging baskets and for air layering (i.e. Spanish moss.).
- **spherical**
  nearly round.
- **spicate**
  arranged in or resembling a spike.
- **spices**
  seeds, fruits, or roots (rhizomes) used to flavour cooking.
- **spiciform**
  spike-like.
- **spiculate**
  having fine fleshy points.
- **spicule**
  male copulatory organ. Sometimes called Spiculum.
- **spider mite**
  any of several small, red, web-spinning mites of the family Tetranychidae that feed on and damage leaves.
- **spike**
  a form of simple inflorescence with the flowers sessile or nearly so upon a more or less elongated common axis.
- **spike harrows**
  a field implement with spikes that breaks up clods and levels soil.
- **spikelet**
  a small or secondary spike, most often referring to those forming the inflorescence of grasses.

- **spiking**
  a method of turf cultivation in which solid tines or flat, pointed blades penetrate the turf and soil surface.
- **spindle-shaped**
  same as fusiform.

- **spines**
a sharp-pointed woody structure, usually a modified leaf or leaf part.

- **spinescent**
1. having spines.
2. terminating in a spine.
3. modified to form a spine.

- **spinule**
a minute spine or spine-like process.

- **spired**
having a crown with several points.

- **spirillum**
1. bacterium with a spiral shape which is relatively rigid.
2. bacterium in the genus Spirillum.

- **spiroplasma**
a member of a group of pleomorphic, wall-less prokaryotes occurring as epiphytes or as intracellular or extra cellular parasites or pathogens in a range of invertebrates and plants.

- **spit**
the depth of one spade's blade.

- **splitting maul**
an axe with a broad, heavy head, shaped into a wide non-stick wedge.

- **spoils**
soil from stream or drain bottoms dredged out when the channel was made deeper.

- **spoon, coring**
a method of turf cultivation involving curved, hollow, spoon like tines that remove small soil cores and leave openings in the sod.

- **sporangiophore**
a modified hypha that supports the sporangium.

- **sporangiospore**
spore formed within a sporangium by fungi in the phylum Zygomycota.

- **sporangium**
a tiny globe in which the spores are produced, often applied to the capsule, but by some authors restricted to the spore sac, or inner sac of the capsule containing the spores.

- **spore**
a microscopic reproductive cell of non-flowering plants (i.e. ferns, lichens, mosses, fungi, and algae). Many times we can see spores on the backside of ferns.

- **sporecase**
same as sporangium, the case in which the spores are borne.

- **sporidium**
1. a basidiospore formed by the rust or smut fungi.
2. in smut fungi: a spore formed in germination by repetition.

- **sporobolomyces**
reported to be allergenic.

- **sporocarp**
the fruit cases of certain cryptogams containing sporangia or spores.

- **sporodochium**
a cluster of conidiophores arising from a stroma or mass of hyphae.

- **sporogenesis**
reproduction by means of spores.

- **sporogenous**
describes cells or tissues in which spores are formed.

- **sporophore**
a spore-producing or spore-bearing structure such as a conidiophores, ascocarp, or basidiocarp.

- **sporophyll**
a leaf bearing spores.

- **sporophyte**
1. the spore-bearing part or generation. In mosses, it consists of the seta and capsule and constitutes the so-called fruit. In ferns, the conspicuous plants that bear spores.

- **spororogonium**
the saprophyte or spore-bearing part of the moss.

- **sporothrix**
can cause sporotrichosis.

- **sporotrichum**
reported to be allergenic.

- **sport**
  an abrupt, naturally occurring genetic change resulting in a branch that differs in appearance from the rest of the plant, or, a plant derived by propagation from such a genetically changed branch.

- **sporulate**
  to produce spores.

- **spot**
  a symptom of disease characterised by a limited necrotic area, as on leaves, flowers, and stems.

- **spot sodding**
  the repair of small areas of damaged turf using plugs of small pieces of sod.

- **spread plate**
  method for performing a plate count of micro-organisms. A known amount of a serial dilution is spread over the surface of an agar plate. After growth the number of colony-forming units is counted.

- **spreader**
  a substance added to fungicide or bactericide preparations to improve contact between the spray and the sprayed surface, a surfactant.

- **spreader-sticker**
  substances added to pesticides to make them spread over and stick to a surface more readily.

- **spreading**
  having a horizontally branching habit.

- **sprig**
  a small part of a plant, such as stolons used for propagations, twigs bearing flowers, etc.

- **spring green-up**
  the initial seasonal appearance of green shoots as spring temperature and moisture conditions become favourable, thus breaking winter dormancy.

- **spring wood**
  wood which is formed during the period of rapid growth in the spring, appearing lighter in colour than the wood formed later as the growth slows down.

- **spud**
  a tool shaped like a flat-ended oar that is used to chop a hole through ice for fishing or studying under-

water biology, but which may also be used to clear ice from sidewalks, cut sod or roots, edge sidewalks, etc.

- **spur**
  a hollow sac-like or tubular extension of some part of a blossom, usually nectariferous.

- **spur branch**
  a stubby branchlet with densely crowded leaves and leaf scars.

- **spurred**
  calcarate, refers to a flower with tubular projections from the petals or sepals, usually with nectar glands.

- **sqarulose**
  diminutively squarrose.

- **squamate**
  covered with scales.

- **squamous**
  covered with or consisting of scales.

- **squamule**
  a small scale or lobe.

- **square-foot gardening**
  a system of gardening developed by Mel Bartholomew that uses 4 foot by 4 foot plots subdivided into 1-foot squares for growing a specific number of a particular type of vegetable to maximise space and facilitate ease of maintenance.

- **squarrose**
  turned back at a right angle or more sharply, as the tip of a phyllary.

- **stability**
  resistance to variation in populations and disturbances, resulting in the persistence of community composition over long periods of time.

- **stachybotrys**
  several strains of this fungus (S. atra, S. chartarum and S. alternans are synonymous) may produce a trichothecene mycotoxin-Satratoxin H - which is poisonous by inhalation. The toxins are present on the fungal spores. This is a slow growing fungus on media. It does not compete well with other rapidly growing fungi.

- **staddle**
  a foundation of trunk and main branches, either of rootstock or stem builder, for grafting.

• **stake**
a piece of pointed wood or metal that is driven into the ground to support a plant.

• **staking**
plants that grow tall with little stem support need to be staked. Perennials and tomatoes come to mind. Any means of support will work: canes, cages, wire loops, etc.

• **stalk**
a short, supporting axis.

• **stalk diameter**
the diameter of a stalk, usually at a designated node or internodes.

• **stalk tunnelling**
longitudinal tunnels in plant stalks produced by insects (e.g., by sugar cane borer, lesser corn stalk borer).

• **stalked**
joined to an axis with a stalk.

• **stamen**
stamens or stamina: The male reproductive organ of a flower. In orchids the one or two stamens are part of the column.

• **staminode**
a sterile stamen, or any structure without anther corresponding to a stamen.

• **staminoid**
a pollen less stamen.

• **standard**
1. a plant with a strong vertical stem, particularly one that would not normally grow that way, a vine or shrub trained as a tree with a single trunk and a more or less spherical top.
2. uppermost, usually erect, petals of some flowers, the upper dilated petal of a papilionaceous corolla.

• **standard atmosphere**
in textile testing: Air maintained at a relative humidity of 65% and a temperature of 70° F (21 to 1° C).

• **staple**
1. natural or cut fibres or cut lengths from filaments.

2. to classify cotton fibre samples by length.

- **starch granule**
  the fundamental unit in which starch is deposited in the storage tissue of many higher plants. It is paracrystalline cold-water insoluble, with a characteristic size and shape depending on the species of the plant that produced it.

- **stationary phase**
  period during the growth cycle of a population in which growth rate equals the death rate.

- **staurospore**
  a non-septate spore with more than one axis.

- **staurosporous**
  possessing staurospores.

- **stegocarpous**
  having the capsule operculate.

- **stele**
  the axial vascular and associated tissues from the endodermis inward.

- **stelidia**
  column teeth.

- **stellate**
  star-shaped.

- **stem**
  the main trunk of a plant. It develops buds and shoots.

- **stem cutting**
  a portion of a stem that only includes one or more nodes taken from a plant. This will not include the apex or the tip. Stem cuttings are a great way to propagate.

- **stem propagation**
  small plants that are formed on flower stems. In some orchids the flower stem has nodes, which carry the dormant eyes and can develop into buds or leaves. These new plants are called 'keiki'. There is a hormone compound called keiki paste that is used in the development of these plantlets. This practice is commonly used on Phalaenopsis.

- **stem tip cutting**
  a cutting taken from the growing tip of a shoot.

- **stem builder**
  a variety of woody plants used in a graft as intermediate stem piece to provide a strong trunk for standard trees, or to introduce resistance to disease or winter injury.

- **stem phylium**
  reported to be allergenic. Isolated from dead plants and cellulose materials.

- **stem-pitting**
  a symptom of some viral diseases characterised by depressions on the stem of the plant.

- **steppe**
  a vast arid, usually level and grassland, particularly those of south-eastern Europe and Asia and generally having extreme variations in temperature.

- **sterigma**
  a spike-like structure that supports a basidiospore on the basidium. (pl. sterigmata).

- **sterigmata**
  the small woody stem to which the leaves of spruces and hemlocks are attached.

- **sterile**
  1. barren. Refers to fern leaves that do not produce spores.
  2. unproductive, as a flower without a pistil, or a stamen without an anther.

- **sterile fungus**
  a fungus that is not known to produce any kind of spores.

- **sterilisation**
  the elimination of pathogens and other living organisms from soil, containers, etc. by means of heat, chemicals, or radiation.

- **sterilised**
  free from living micro-organisms

- **sterilised soil**
  it is soil that is steam- or chemically sterilised. Harmful organisms have been killed but helpful bacteria have been spared.

- **stichidium**
  a small structure (lobe or branch) bearing tetra sporangia.

- **sticker**
  added to fungicide or bactericide preparations to improve the adhesion of the spray to the sprayed surface

- **stigma**
  pl. stigmas or stigmata The terminal part of the ovary, at the end of the style, which is receptive to the pollen.

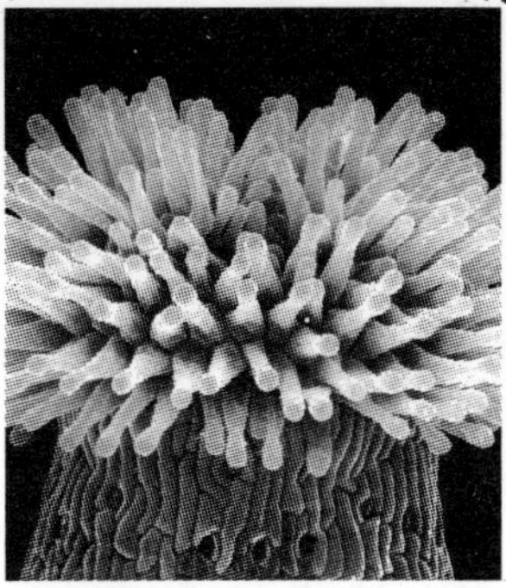

- **stigmatic**
belonging to or characteristic of the stigma.

- **stigmatose**
with conspicuous stamens.

- **stigonema**
a filamentous blue-green algae.

- **stilt-roots**
adventitious support roots.

- **stimulants**
herbal medicines that can increase the energy or activity of a tissue, organ, etc.

- **stipe**
1. the stalk-like support of a pistil.
2. the petiole of a fern or palm frond.
3. the stalk supporting the fruiting body of some fungi.

- **stipel**
a stipule of a leaflet.

- **stipule**
an appendage at the base of a petiole, often appearing in pairs, one on each side, as found on roses.

- **stipule scar**
the scar left on a twig by the fall of a stipule.

- **stock**
the 'mother plant' of which cuttings are taken. Stock is also in reference to plants being grafted on 'stock'. Many hybrids are grafted on good stock material because they have better and sturdier qualities for growing.

- **stock driveway**
a lane, a strip of land designated for the movement of livestock.

- **stock plant**
a plant used as a source for cuttings.

- **stock seed**
supply of seed, tubers, or roots reserved for planting and used as a source of germ plasma for maintaining and increasing seed of crop varieties or genetic lines. Synonym: **seed stocks**.

- **stocking density**
  the relationship between the number of animals and the specific unit of land being grazed at a given time.

- **stocking plan**
  the number and kind of livestock assigned to one or more given management areas or units for a special period.

- **stocking rate**
  the relationship between the number of animals and the grazing management unit.

- **stolon**
  sometimes confused with runner, this horizontally growing stem roots at its tip, not its nodes.

- **stolon nursery**
  an area used for producing stolons for propagation.

- **stoloniferous stem**
  a slender creeping stem with minute leaves.

- **stoloniform**
  like a stolon, underground stem.

- **stoma**
  1. an orifice in the epidermis of a leaf communicating with internal air cavities.
  2. a pore in the wall of a capsule surrounded by special guard-cells and serving the same purpose as the stomata in the epidermis of the leaves of flowering plants.

- **stomata**
  an opening or pore in leaves that is surrounded by guard cells.

- **stone**
  the single seed of a drupe, surrounded by a large, hard shell and covered by pulp.

- **stool**
  clump of shoots growing from near the ground.

- **stooling**
  cutting down to ground level to induce tightly packed new growth.

- **stopping**
  pinching off the terminal bud to induce branching.

- **storage**
reversible accumulation of materials in localised organelles, cells, or organs. Storage implies that the material is available for subsequent mobilisation under some conditions.

- **storage polysaccharide**
energy reserve deposited in the cell when there is an excess of carbon available. These are usually deposited as large granules in the cell. The most common example of a storage polysaccharide in plants is starch. Its counterpart in animal cells is glycogen.

- **storm surge**
before and during a storm, an abnormal and rapid rise in lake level along the shore, caused mostly by strong onshore winds and changes of atmospheric pressure.

- **stove**
a heated greenhouse.

- **stove plant**
a plant which requires warm greenhouse conditions in winter.

- **straggling**
semi upright.

- **straight**
describing ray florets having little or no curvature throughout their length (used in describing dahlias.).

- **strain**
a number of plants which have similar characteristics such as colour blends, different enough to be designated, but not sufficiently distinct to be considered a horticultural variety.

- **stramineous**
straw-coloured.

- **strand**
1. the area of lakeshore above the waterline that is subject to the action of wind and rain.
2. on ocean beaches, the area between high and low tidal marks.

- **stratify**
artificially overcoming a seed's dormancy by placing them in layers of moisture-retaining media and keeping them under generally cool and moist conditions for a period of time, so as to simulate winter conditions.

- **stratigraphy**
  the study of layers of rock in reference to their ages.
- **straw**
  the hollow stem of grass.
- **strawflowers**
  species of plants which easily become dried flowers.
- **strength, breaking**
  the maximum internal cohesive forces of a material that resist rupture during a tensile test.
- **stress**
  an environmental condition potentially harmful to an organism, and causing a reduction in normal functions.
- **stress, tensile**
  the resistance to deformation developed within a specimen subjected to tension by external force.
- **stretching the zone**
  for very ambitions gardeners who want to grow plant material beyond their climate area. Collectors are forever trying to grow what might not be grown.
- **stria**
  a thin ridge or groove.
- **striae**
  in nematodes: Superficial grooves or clefts on the cuticle. When present, stria may be seen encircling lips or body.
- **striate**
  marked with delicate lines, grooves, or ridges.
- **strict**
  very straight and upright.
- **strigose**
  beset with apprised sharp, straight and stiff hairs.
- **strigulose**
  minutely strigose.
- **string tags**
  waterproof tags used to mark collected specimens for herbarium study.
- **striolated**
  with delicate lines or ridges.
- **strip cropping**
  a method of planting a crop in parallel bands following the contours of the slope, this lessens erosion.
- **strip sodding**
  laying of sod strips spaced at intervals, usually across a slope. Turf establishment depends on spreading of the grass to form a complete

cover, sometimes the area between the strips is seeded.

- **strobiloid**
  cone-shaped.

- **strobilus**
  1. woody cone, as in Gymnospermae.
  2. a cone-shaped inflorescence marked by imbricated sporophylls.
  3. a dense cone-like structure producing spores.

- **stroma**
  1. a compact mass of fungal hyphae producing perithecia or pycnidia.
  2. the colourless proteinaceous matrix of a chloroplast in which the chlorophyll-containing lamellae are embedded.

- **strone**
  a hill that terminates a ridge, the end of a ridge.

- **strophile**
  an appendage at the hilum of certain seeds.

- **strophiole**
  resulting from an inverted ovary that has fusion of the nucleus and funiculus. The strophiole is a small swelling on a ridge.

- **structural polysaccharide**
  polysaccharide that serves primarily as a structural element in cell walls and coats and intercellular spaces, and connective tissue where they give shape, elasticity, or rigidity to plant or animal tissues and protection and support to unicellular organisms. Cellulose is the predominant structural polysaccharide in plant cell walls and chitin is abundant in fungal cell walls and insect exoskeletons.

- **structure**
  see **soil structure**.

- **struma**
  a goitre-like swelling on one side at the base of the capsule.

- **stubble mulch**
  crop residues left essentially in place after tillage, as a surface cover during fallow and the growing of a succeeding crop.

- **stunt virus**
  a type of virus disease which stops the growth of a plant.

- **stylet**
  in nematodes: hollow protrusible spear used to

puncture plants or animal prey.

- **stylet knob**
one of the (usually three) basal protuberances of the stylet.

- **stylopodium**
a disc-shaped enlargement found at the base of the style in some Umbelliferae.

- **stylospore**
a spore borne on a filament or hypha.

- **styptic**
a medicine to stop minor surface bleeding, e.g., razor cuts.

- **subarctic**
of, relating to, or being the regions bordering on the arctic zone.

- **suberect**
ascending toward the edges.

- **suberin**
a complex fatty substance found especially in the cell walls of cork.

- **suberised**
of cell walls: Hardened by their conversion to cork

- **suberous**
cork-like.

- **subgenus**
one of the divisions into which large genera are sometimes taxonomically divided.

- **subimbricate**
somewhat overlapped, slightly shingled.

- **submergent vegetation**
plants with stems and leaves below water level, sometimes with reproductive parts above water.

- **submersed**
underwater.

- **subopposite**
almost opposite but one leaf or leaflet of each pair a little above the other.

- **subsessile**
with a slight stalk.

- **subshrub**
a plant with more or less woody branches and roots, less woody than a true shrub, but more woody than a perennial herbaceous plant, a small shrub, woody only at the base.

- **subsoil**
the usually less fertile soil below the topsoil.

- **subspecies**
  the rank of taxon below species but above variety, a subdivision of a species whose members have certain hereditary characteristics distinct from other populations of that species. A subspecies is added to the specific binomial and preceded by 'ssp.,' such as obtusifolia in the epithet Grevillea thelemanniana ssp. obtusifolia.

- **substrate**
  1. substance, base, or nutrient on which an organism grows.
  2. compounds or substances that are acted upon by enzymes or catalysts and changed to other compounds in the chemical reaction.

- **substrate-level phosphorylation**
  synthesis of high-energy phosphate bonds through reaction of inorganic phosphate with an activated (usually) organic substrate.

- **substratum**
  the material on which a plant grows.

- **subtend**
  to be just below and close up to or enclosed in its axil.

- **subterranean**
  below ground.

- **subtropical**
  very specific area, 5 to 10 degrees higher in latitude than the Tropic of Cancer or the Tropic of Capricorn.

- **subulate**
  awl-shaped.

- **succession**
  1. the development of a plant community from its initial stage to its climax stage, usually from one consisting of grasses and forbs to one of shrubs and, finally, to forest. However, depending on the climate, the climax stage may consist of a prairie, savana, or any number of other ecosystems.
  2. the changes in the species composition of communities following a natural or human disturbance like the natural filling of a pond or the clearing of a road through a forest.

- **succession planting**
  when one plants a fast crop one week or so after an-

other. The object is to keep a constant supply on hand, like squash, lettuce. Truck farmers often practice this technique.

- **succulent**

succulent plants have leaves and/or stems which are thick and fleshy. They often have waxy outer layers that allow the plants to retain water well.

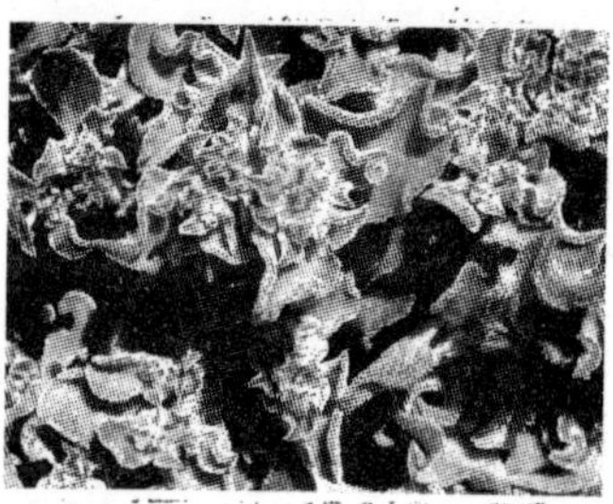

- **sucker**

a shoot from the root or lower part of a stem. In roses, a young cane emerging below the bud union and therefore representing the variety of the under stock rather than the top variety. A shoot appearing on a tree limb is called a water sprout.

- **suffrutescent**

slightly or obscurely shrubby.

- **suffrutex**

subshrub.

- **suffruticose**

very low and woody, diminutively shrubby.

- **sulphur cycle**

sequence of transformations undergone by sulphur where it is taken up by living organisms, transformed upon death and decomposition of the organism, and converted ultimately to its original state of oxidation.

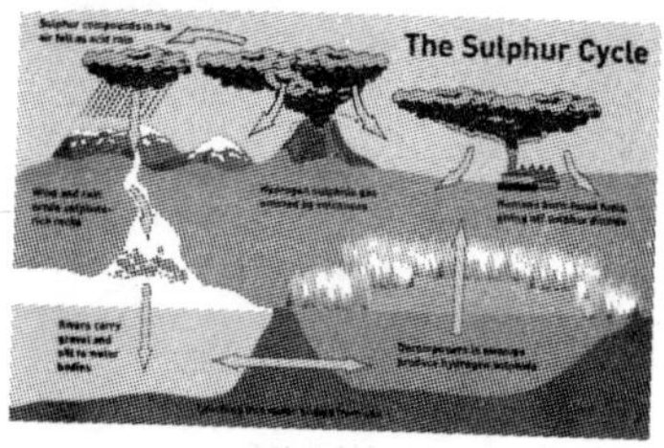

- **summer syndrome**

usage: An unsatisfactory term for fescue toxicosis, since it implies a seasonal restriction. Acceptable if limited to signs observed in the summer. Summer slump is a vernacular synonym, also unsatisfactory.

- **summer wood**

wood produced late in the growing season when growth is slow.

- **sun scorch**

spots on leaves that are caused by exposure to

strong sunlight. Often not acclimating plants for the season creates sunspots. Just trim off and let new growth develop.

- **sunken garden**
  a garden recessed into the ground to create a secluded atmosphere or to facilitate a view down onto a feature such as a knot garden.

- **sunscald**
  plant injury caused by exposure to bright sunlight, excessive heat and/or wind.

- **superficial**
  on the surface.

- **super infection**
  attempt to infect a host with a second virus, usually a different strain of the first infecting virus

- **superior ovary**
  with the flower parts growing from below the ovary.

- **supernatant**
  the material remaining above the pellet after centrifugation of a suspension.

- **superparasitism**
  parasitisation of a host by more than one individual usually of a single species.

- **superphosphate**
  a fertiliser with a very high amount of phosphates usually 30-45 percent, and most often used to promote flowering in ornamental plants.

- **superposed**
  accessory buds which occur above the axillary buds.

- **supersaturation**
  the uptake of unusual amounts of water by a plant, this can cause plant specimens to mould rather than dry.

- **supersporangium**
  a structure containing tetra sporangia and acting as a sporangium.

- **supplemental feeding**
  of animal feeding: supplying concentrates or harvested feed to correct deficiencies of the pasture diet. Often erroneously used to mean emergency feeding.

- **supplements**
  in nematodes: pre-anal genital papillae on the ventral side of males, derived from cuticle, but may be provided with glands.

- **suppression**
  a hypoplastic symptom characterised by the failure of plant organs or substances to develop.
- **suppressive soil**
  a soil in which certain disease(s) fail to develop because of the presence in the soil of micro-organisms antagonistic to the pathogen.
- **supra-axillary**
  borne above the axil.
- **suprabasal**
  above the base, usually referring to pinnae or pinnules.
- **supramedial**
  beyond the middle.
- **surculose**
  producing suckers.
- **surcurrent**
  extending upward on and adnate to an axis.
- **surface area**
  area of the solid particles in a given quantity of soil or porous medium.
- **surface drainage**
  grading or piping that allows excess water to be carried away by gravity.
- **surface soil**
  uppermost part of the soil, ordinarily moved in tillage, or its equivalent in uncultivated soils ranging in depth from 7 to 20 cm. Frequently designated as the surface layer, the Ap layer, or the Ap horizon.
- **suscept**
  any plant that can be attacked by a given pathogen, a host.
- **susceptibility**
  the inability of a plant to resist the effect of a pathogen or other damaging factor.
- **susceptible**
  lacking the inherent ability to resist disease or attack by a given pathogen, not immune.
- **susceptible host**
  an organism that can be infected by a pathogen.
- **suspended**
  describes an ovule that is hanging from the apex of the cell.
- **suspensor**
  a specialised hyphal tip that supports a gametangium

and eventually a zygo sporangium (in the Mucorales).

- **sustainability**
  the capability of entities to survive without human intervention.
- **sustainable development**
  human intervention that meets the needs and aspirations of the current generation without compromising the ability to meet those of future generations.
- **sustained yield**
  the continuation of desired forage or animal production.
- **suture**
  a line of dehiscence.
- **swale**
  a wet depression between beach ridges, fed by ground water and runoff, rarely influenced by changes in water level.
- **swamp**
  an area of land that is usually flooded and contains woody plants. Compared to a bog it has drainage. Cypress trees grow well in a swamp.
- **swarmspore**
  zoospore.
- **swash marks**
  concentrated debris above the waterline and parallel to it covered and uncovered by uprushing waters.
- **swash zone**
  the shore area above the waterline, covered and uncovered by uprushing water.
- **swathing**
  the cutting or pulling of a crop into a windrow to dry and/or to facilitate harvest.
- **sweating**
  the process of placing plant specimens in a plant press for 24-48 hours, then opening the press. Moisture will have been drawn out of the plant into the driers but plant parts will still be pliable enough to make final arrangements of parts.
- **swell**
  a huge natural dome structure where movement of the earth's crust has raised stratigraphic layers in mounds rather than breaking them with a fault. Erosion exposes the oldest lay-

ers in the middle with subsequent layers surrounding it in concentric circles.

- **swelling clay**
  a soil with little sand which shrinks when dry and swells when wet. This recurring problem usually strips bark from roots of woody plants, so the area favours grasses and other non-woody species.

- **swidden**
  a temporary agricultural plot produced by cutting back and burning the current plant life.

- **swoe**
  a type of scuffle hoe.

- **syconium**
  a fleshy, multiple fruit having a hollow centre, like that of a fig.

- **sylvan**
  of, or relating to, a forest or woods.

- **symbiont**
  one member of a symbiotic relationship.

- **symbiosis**
  the living together of different species of organisms which may or may not be to their mutual benefit.

- **symbiotic**
  a relationship in which two or more dissimilar organisms live together in close association.

- **symmetrical**
  describes a flower that can be divided into similar halves.

- **sympatric**
  describes two or more closely-related species having coincident or overlapping ranges of distribution but not interbreeding.

- **sympetalous**
  gamopetalous, with petals joined at the margins or at the base, often forming a tube or funnel.

- **sympodial**
  a specialised growth pattern in which the terminal bud dies or blows in an inflorescence with growth continuing by development of more or less equal side branches.

- **sympodium**
  orchids that produce new growth from the base of the plant where the previous growth occurred (from the

rhizome). The majority of orchids have sympodial growth; the others have a monopodial growth pattern.

- **symptom**
a visible abnormality in a plant that results from disease.

- **symptomatology**
the study of symptoms of disease and signs of pathogens for the purpose of diagnosis

- **symptomless carrier**
a plant that, although infected with a pathogen (usually a virus), produces no obvious symptoms.

- **synanamorph**
any one of the two or more anamorphs that are formed by the same fungus.

- **synandrium**
with the androecium having the anthers connate.

- **synangium**
an oblong group of fused sporangia.

- **synantherous**
describes a plant which leafs out and flowers at the same time.

- **synanthous**
refers to leaves which open at the same time as the flowers.

- **syncarp**
an aggregate or multiple fruit produced from fused or coherent pistils, the small individual fruits massing and growing together into a single fruit.

- **syncarpous**
with the carpals of the gynaecium united in a compound ovary.

- **syncephalastrum**
can cause a respiratory infection characterised by a solid intricately fungal ball.

- **syncline**
a geological structure that appears when strata fold downward toward a common line.

- **syncytium**
a multinucleate cell.

- **synergid**
  one of two small cells lying near the micro pyle of the embryo sac of an angiosperm.

- **synergism**
  the concurrent parasitism of a host by two pathogens in which the symptoms or other effects produced are of greater magnitude than the sum of the effects of each pathogen acting alone.

- **synflorescence**
  a compound inflorescence composed of both terminal and lateral flowering branches.

- **syngenesia**
  plants with the anthers fused, forming a ring around the style, e.g., Compositae.

- **synnema**
  a fascicle of conidiophores, usually upright, a coremium.

- **synoicous**
  having the male and the female organs mixed together in the same cluster.

- **synonomy**
  historical listing of all the names applied by different authors to a plant.

- **synonym**
  any one of two or more names used for the same taxon rank, a rejected name due to misapplication or difference in taxonomic judgment.

- **synsepal**
  a single floral part formed by the partial or complete fusion of two or more of the orchid sepals (usually the lateral pair as in a Paphiopedilum).

- **synsepalum**
  a cup or tube made up of two or more united sepals, a fused calyx.

- **synthetic**
  substances produced by chemical or biochemical means.

- **syntrophy**
  interaction of two or more populations that supply

each other's nutritional needs.

- **syntype**
  one of a number of specimens of equal nomenclature rank that formed all or part of the material the original author had in cases in which the author did not designate or indicate a holotype.

- **syringing**
  spraying turf with small amounts of water to dissipate accumulated energy in the leaves by evaporating free surface water, prevent or correct a leaf water deficit, particularly wilt, and remove dew, frost, and exudates from the turf surface.

- **systemic**
  anything introduced into the growth system of a plant and travelling throughout the system. Some types of systemic are used to protect plants, as a systemic insecticide. Other systemic are used as herbicides to kill plants deemed as noxious weeds.

- **systemic infection**
  in virology, an infection resulting from the spread of virus from the site of infection to all or most cells of an organism.

- **systemic insecticide**
  a pesticide which can be granular or liquid, used at the base of the plant and travels through the vascular stream.

- **systemics**
  scientific study of the kinds and diversity of organisms and of the relationships between them.

- **systilius**
  the lid continuing fixed to the columella, and thus elevated above the capsule when dry.

- **tail**
  in nematodes: the portion of the body between the anus and the posterior terminus.

- **tall fescue**
  a coarse, clump-forming turf grass that tolerates low fertility, heavy wear, heat and drought and has good insect and disease tolerance. Young seedlings are not cold tolerant, but mature plants survive most winters.

• **tamping**
a method by which one will press that soil around a plant that has just been planted making sure soil is secure and firm around the roots.

• **tanbark**
residue of bark that has been used for tanning leather, useful for mulch or pathways.

• **tannin**
any of various soluble astringent complex phenolic substances of plant origin, possibly giving protection to the plants. Tannins are used especially in tanning, dyeing, and making inks and pharmaceuticals.

• **tapered**
gradually narrower or thinner.

• **tapetum**
a layer of nutritive cells that invests the sporogenous tissue in the sporangium of higher plants.

• **taproot**
the main root of a plant, having a single, dominant axis and often surviving the functions of structural support and food storage.

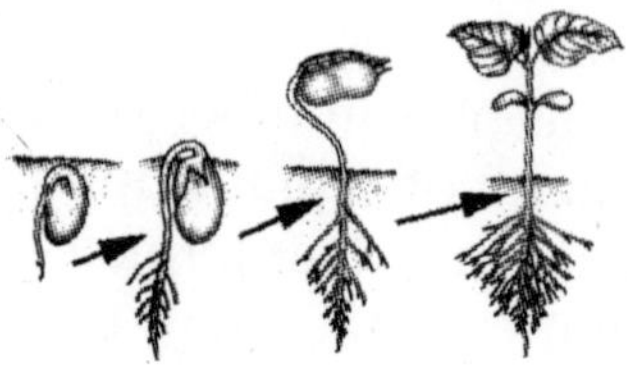

• **tarsonemid mites**
small mites living and feeding on young plant tissues.

• **tautonym**
an illegitimate binomial in which the genus and species are the same word, such as Armoracia armoracia, later changed to Armoracia rusticana.

• **taxon**
a group of genetically similar organisms that are classified together as a species, genus, family, etc.

• **taxonomic synonym**
any one of two or more different names based on different plant specimens which were later judged to belong to the same taxon.

• **taxonomy**
study of scientific classification and nomenclature.

- **teichoic acids**
  all wall, membrane, or capsular polymers containing glycerophosphate or ribitol phosphate residues.

- **telamon**
  in nematodes: rigid, sclerotised portion of the cloacal wall that apparently guides the spicules from the spicular pouch into the cloaca.

- **teleomorph**
  the stage characterised by the production of asci/ascospores, basidia/basidiospores, teliospores, or other basidium-bearing organs.

- **teleutosorus**
  an old term for telium.

- **teleutospore**
  see **teliospore**.

- **teliospore**
  a usually thick walled, usually resting spore, which is the site of karyogamy and produces the basidium (in the Uredinales and Ustilaginales).

- **telium**
  the final sorus (stage III) produced in the life cycle of the Uredinales and producing teliospores. (pl. Telia).

- **temperate**
  moderate, the zones between the tropics and the polar regions of the earth.

- **temperate virus**
  virus which upon infection of a host does not necessarily cause lyses but whose genome may replicate in synchrony with that of the host.

- **temperature summation**
  the totality of effective temperatures or day-degrees for a specified period of time, especially the time required for an organism to reach maturity, like a summer.

- **tempered**
  brought to predetermined moisture characteristics or temperature (or both) before further processing.

- **template organism**
  the original when genetic engineering is being practised on a species.

- **temporal**
  1. refers to the sides of the skull behind the eyes, the temples.

2. refers to time.

- **tenacity**
  the tensile stress expressed as force per unit linear density of the unstrained specimen (e.g., of cotton).
- **tender**
  an indoor plant which requires a minimum temperature of 60° F. Occasional short exposure to temperatures below this level may be tolerated.
- **tendril**
  plants (like sweet peas, clematis, and grapes) produce a cordlike structure that will help to support them.

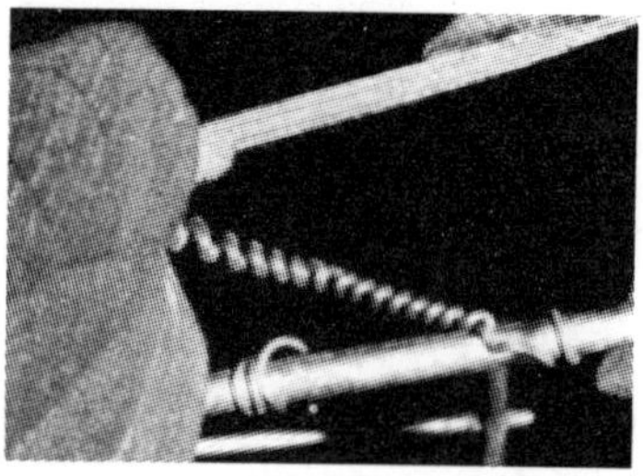

- **tension**
  the stress resulting from elongation.
- **tepal**
  a division of the perianth, a sepal or a petal of a flower in which the calyx and corolla are almost identical in appearances, as occurs in tulips and lilies. In Orchidaceae, this term refers to any sepal or petal, except the labellum.
- **tepee**
  a tripod of stakes used to support climbing plants.
- **teratological**
  monstrous, relating to a monstrosity.
- **terete**
  having a circular transverse section.
- **terminal**
  the uppermost bud or flower on a stem.
- **terminal bud**
  a bud formed at the tip of a stem, twig, or branchlet.
- **terminal electron acceptor**
  external oxidant (often oxygen) that accepts the electrons as they exit from the electron transport chain.
- **ternate**
  with three nearly equal divisions.
- **terra cotta**
  an Italian term that means 'baked earth'. These clay pots are unglazed and excellent for growing most plant

material. They do dry out quickly and salts will bleed through the porous surfaces.

- **terrace**
  a series of flat platforms of soil on the side of a hill, rising one above the other.

- **terrarium**
  a small artificial environment for a specified habitat. An aquarium is for a water habitat, usually containing fish and plants. A terrarium generally has a woodland setting, with the emphasis on plants rather than animals.

- **terrestrial**
  plants that grow in or near the surface of the ground, growing in soil. Orchids generally are found growing one of three ways: as terrestrials

- **terrigenous**
  refers to deposits from land sources.

- **tessellate**
  in nematodes: checkered, a type of cuticular pattern in which the longitudinal ridges are broken by transverse striations into rows of squares.

- **test**
  hard external covering or shell.

- **tester animals**
  animals of like kind and similar physiological condition used in grazing experiments to measure animal performance or pasture quality, usually assigned to a treatment for the duration of the grazing season, as opposed to grazer animals that may be assigned temporarily.

- **teteur**
  a French word meaning 'trainer'. It is a structure that is in the shape of an obelisk or pointed tower. This trellis will support vines of colour.

- **tetrad**
  a group of four.

- **tetradynamous**
  having four long and two shorter stamens.

- **tetragonal**
  four-angled.

- **tetramerous**
  having flower parts (petals, sepals, stamens) in groups of four.

- **tetraploid**
  having twice the normal number of chromosomes.
- **tetrapterous**
  with four wings.
- **tetrasporangium**
  the sporangium (meiotangium) containing four spores.
- **tetraspore**
  one of the four spores in a tetrasporangium.
- **texture**
  1. the surface, appearance, and hand of a textile material.
  2. in turf, leaf texture refers to leaf width, taper, and arrangement.
- **texture**
  the arrangement of the components of the different tissues, as compact, loose, etc
- **thallic**
  one of two basic kinds of conidiogenesis, the conidial initial enlarges after it has been delimited by one or more septa. The conidium is differentiated from a whole cell.
- **thalloid**
  resembles a thallus.
- **thallophyta**
  a primary taxonomic division of organisms which are plant-like, but not true plants, including all one-celled and cell aggregates which do not clearly show root, stem, and leaf. Includes algae, fungi, and lichens.
- **thallophyte**
  1. any plant or plant-like organism belonging to the division Thallophyta with single-celled sex organs or with many-celled sex organs of which all cells give rise to gametes, including the algae, fungi, and lichens
  2. a plant that absorbs its food supply over its growing surface, as a seaweed or fungus mycelium.
- **thallospore**
  a conidium that has no conidiophore or is not separate from the hypha or conidiophore that produced it
- **thallus**
  in cryptogams, a cellular expansion taking the place of stem and foliage and

forming the main body fungi and lichens.

- **thatch**
any material that does not quickly decompose. When a lawn becomes 'clogged' with old drying grass cuttings and matted leaves, this is referred to a build up of thatch. Often it does not decompose fast enough for a healthy lawn and needs to be removed, either manually or by a thatching machine.

- **thatch control**
preventing excessive thatch accumulation by cultural manipulation and/or reducing excess thatch by mechanical or biological means.

- **theca**
a case, usually referring to the pollen sac in flowering plants or the capsule in bryophytes.

- **therapy**
principle of plant disease control marked by the cure of disease, as with heat or systemic chemicals.

- **thermal stratification**
the successive horizontal layers of a body of water having different temperatures, each layer more or less sharply different from the adjoining ones, with the warmest at the top.

- **thermocline**
in bodies of water, there are normally stratified layers of decreasing temperature with increased depth. Thermocline is the layer where the temperature starts to decrease rapidly, about one degree centigrade or more with each meter (3 feet) of depth.

- **thermophile**
organism whose optimum temperature for growth is between 45° C and 85° C.

- **thermophotoperiodic**
describes growth affected by both temperature and exposure to light.

- **thicket**
any area that has a lot of miscellaneous undergrowth, generally of small shrubs, bushes, and vines.

- **thigmotropism**
a tropism in which contact especially with a solid or a rigid surface is the orienting factor.

- **thin**
to reduce the number of seedlings or young plants to give the remainder room to grow.

- **thinning**
picking out the overpopulated seedlings in any flower or vegetable bed, to make a better growing condition for the rest. Making better spacing and aesthetics for the growing area.

- **thinning out**
1. the removing of all interior branches of a shrub or tree to allow better air circulation.
2. the removing of excess seedlings to allow stronger growth of the remaining ones.

- **thorax**
the middle of the three major divisions of the arthropod body.

- **thorn**
a stout, sharp, woody outgrowth of the stem or branch.

- **threadworm**
nematode

- **thrice-compound**
divided into leaflets that in turn are divided into leaflets, and then further divided into sub leaflets.

- **thrips**
insects that feed on all parts of the plant: leaves, flowers, buds, and stems. Very popular in the destruction of daylilies (this word is both singular and plural).

- **throat**
the orifice of a gamopetalous corolla or calyx, the part between the proper tube and the limb.

- **thug**
not your normal street gangster, but close to it. Thugs are invasive plants. We enjoy their attractiveness but they will take over

a garden. They are growing in the right conditions, but ironically thugs can make a statement when they are perennials in the right garden.

- **thylakoid**
  any of the membranous disks of lamellae within plant chloroplasts that make up the grana and are the sites of the photochemical reactions of photosynthesis.

- **thyrse**
  a contracted cylindrical or ovoid and usually compact panicle.

- **thyrsiform**
  similar to a thyrse.

- **thyrsus**
  the main stem of an inflorescence, indeterminate and paniculate or racemose, but with side branches which are determinate and cymose or dichasially compound.

- **ti plasmid**
  conjugative tumour-inducing plasmid present in the bacterium. Agrobacterium tunefaciens which can transfer genes into plants.

- **tidal pool**
  a small pool of water which at low tide is isolated from a larger body of water. A habitat created by such a pool and the plants and animals found there.

- **tier**
  a radial, layered branching habit of ex-current trees.

- **tif**
  a commercial name for a series of hybridised Bermuda grasses which survive daily cutting, are often used on golf greens, but are far too much work for the home yard.

- **tiling**
  the installation of permeable pipes buried below the surface of a field to improve drainage.

- **till**
  mixed deposits of gravel, boulders, sand, and other materials that are carried by a glacier.

- **tillage**
  for other entries, see also bare fallow, chemical fallow, conventional tillage, minimum tillage, no-till, primary tillage, reduced tillage,

secondary tillage, summer fallow, till-plant, zero till.

- **till-plant**
  seedbed preparation and planting completed in the same operation, or one immediately following the other, leaving a protective cover of crop residue on and mixed in the surface layer. In some areas, referred to as minimum tillage.

- **tilth**
  1. cultivated land.
  2. the physical condition of soil and its suitability for planting.

- **timber**
  standing trees, felled trees, or seasoned logs.

- **timber cruiser**
  a person hired by a wood products company, or a government employee, who locates standing crops of timber and estimate their yield of lumber.

- **tincture**
  an herbal preparation made by soaking plant parts in alcohol.

- **tip**
  the apex.

- **tip cutting**
  a cutting taken from the top end of a shoot.

- **tip layering**
  a propagative method similar to air layering, except a flexible branch such as blackberry, Rubus, is bent to the ground and buried in moist soil or a mixture of peat moss and sand, instead of being wrapped on the upright stem.

- **tired soil**
  a term referring to a piece of land that has been exhausted of its nutrient value. It does not produce like it once used to. When a particular crop has been grown too long in once place.

- **tissue**
  a group of cells of similar structure which performs a special function..

- **tissue culture**
  a very sterile practice of propagating plants from the mother plant. Generally done in laboratory conditions. Orchids, hosta and daylilies are done by this method.

- **titillae**
  in nematodes: small projections on either side of the distal end of the gubernaculum.

- **toadstool**
  mushroom.

- **tobamovirus**
  siglum of tobacco mosaic virus. Member of a group of plant viruses with rigid, rod-shaped particles containing one molecule of linear RNA, easily sap-transmitted and transmitted in nature by contact and (in the case of some viruses in this group in seed).

- **tolerance**
  1. the ability of a plant to sustain the effects of a disease without dying or suffering serious injury or crop loss.
  2. the amount of toxic residue allowable in or on edible plant parts under the law.

- **tolerance range**
  range of environmental conditions in which an organism can survive, set mainly genetically, but modified by previous environmental history of the individual.

- **tolerant carriers**
  plants which carry disease but are not themselves harmed by the disease.

- **tomato hornworm**
  the larva of the hawkmoth, *Manduca quinquemaculata*, which feeds on plants of the nightshade family, especially tomatoes.

- **tombusvirus**
  siglum of tomato bushy stunt virus. Member of a group of plant viruses with isometric (icosahedral) particles containing one molecule of linear RNA, readily sap-transmitted and also transmitted through soil.

- **tomentose**
  covered with a thick felt of radicals, densely pubescent with matted wool.

- **tomentum**
closely-matted woolly hairs.

- **tonics**
herbal medicines that can sharpen the appetite and promote strength.

- **tonoplast**
a semi-permeable membrane surrounding a plant cell vacuole.

- **tonsure**
the shaping of evergreens by clipping.

- **tooth**
1. the processes comprising the fringe around a peristome.
2. the smallest division of a frond.

- **top cross**
the first generation of a cross between an open-pollinated variety and an inbred line, a foundation backcross, or a foundation single cross.

- **top variety**
the variety bud-grafted to the under stock, and thus the variety that flowers.

- **top water**
to pour water into the top of a houseplant or other container. This may be merely enough to replenish moisture, or may be in copious quantities to dissolve and carry out salt build-up on the top of the soil.

- **topdressing**
a prepared soil mix added to the turf surface and worked in by brooming, matting, raking, and/or irrigating to smooth a green surface, firm a turf by working soil in decomposition, cover stolons or sprigs during vegetative decomposition, and cover stolons or sprigs during vegetative planting.

- **topiary**
the art of clipping and training woody plants to form geometric shapes or intricate patterns. Box and Myrtle are suitable types.

- **topography**
the configuration of a surface including its relief and the position of its natural and man-made features.

- **topotype**
a specimen collected later from the original type locality, or from the area

from which the type was described.

- **topsoil**
  1. layer of soil moved in cultivation.
  2. the horizon.
  3. presumably fertile soil material used to top dress road banks, gardens, and lawns.

- **torose**
  cylindrical with contractions at intervals.

- **torrent**
  a violent stream, a flood.

- **tortuous**
  bent or twisted in different directions.

- **torula**
  reported to be allergenic.

- **torulose**
  diminutive of torose.

- **tospovirus**
  siglum of tomato spotted wilt virus. Member of a small group of plant viruses with spherical particles containing three RNA species, transmitted in nature by thrips in a persistent manner

- **total digestible nutrients**
  sum total of the digestibility of the organic components of plant material and/or seed (e.g., crude protein + nitrogen-free extract + crude fibre + fat).

- **totipotency**
  the ability to generate or regenerate a whole organism from a single cell.

- **totipotential cells**
  a cell capable of regenerating an entire organism.

- **toughness**
  that property of a material by virtue of which it can absorb work.

- **toxicity**
  1. being poisonous or toxic.
  2. injury, impairment, or death resulting from a poison or toxin, i.e., a toxic reaction. Synonym: toxicosis.

- **toxicology**
  the study of poisons.

- **toxicosis**
  a condition resulting from poisoning.

- **toxigenic**
  producing poison.

- **toxin**
  a compound produced by a micro-organism and being toxic to a plant or animal.
- **toxoid**
  toxin that has been treated to make it non-toxic, but that will still induce the formation of antibodies.
- **trabeculate**
  describes peristome teeth with prominent transverse bars.
- **trace elements**
  tiny quantities of chemical elements.
- **trace gas**
  gas other than nitrogen and oxygen in the atmosphere, particularly those gases that are active in the chemistry or radiation balance of the atmosphere.
- **tracheid**
  a lignified water-conducting cell.
- **tracheophyta**
  a major taxonomic division of plants containing green plants with tubular food and water transport systems: Pteridophyta and Spermatophyta, commonly called vascular plants.
- **trade name**
  an arbitrary name created by a nursery or some other organisation to distinguish it from all others. A way of protecting it from misuse legally.
- **trailing**
  any plant that grows long stems and will grow along the ground and will root as it goes along.
- **trama**
  the sterile tissue of a basidiocarp
- **transcription**
  synthesis of an RNA molecule complementary to one of the two strands of a DNA double-stranded molecule.
- **transduction**
  the virus-mediated transfer of host DNA (chromosomal or plasmid) from one host cell (the donor) to another (the recipient). Transduction was first observed in bacteriophage/bacterium systems, but has since also been found to be mediated by certain vi-

ruses infecting eukaryotic cells.

- **transfection**
  the successful virus-infection of cells following their innoculation with viral nucleic acid.

- **transfer RNA (tRNA)**
  type of RNA that carries amino acids to the ribosome during translation.

- **transformant**
  a cell or organism that has undergone genetic transformation.

- **transformation**
  a process in which exogenous DNA is taken up by a recipient cell or protoplast, in which it may be incorporated into the chromosome or, e.g., into a plasmid by homologous recombination or converted into an autonomous replicon. The DNA (transforming or donor DNA) may be a fragment of chromosomal DNA from a related strain, a plasmid, or a viral genome.

- **transgenic**
  describes genetically modified plants or animals containing foreign genes inserted by means of recombinant DNA techniques.

- **transgressive beach**
  a new beach formed during a period of high water, destroying an old beach and making a new barrier further inland.

- **transit**
  a surveyor's instrument for determining position and elevation.

- **transitional climatic zone**
  the sub-optimal zone between the cool and warm climates, where both warm- and cool-season grasses can be grown.

- **translation**
  synthesis of proteins using the genetic information in mRNA as a template.

- **translocation**
  transfer of nutrients or virus through the plant.

- **translucent**
  thin enough to transmit light, but not a concrete image.

- **transmission**
  the transfer of a pathogen from one plant to another,

or from one plant organ to another.

- **transpiration**
  the loss of water vapour from the surface of leaves and other aboveground parts of plants.

- **transplant**
  to move a plant from one place to another.

- **transposable element**
  genetic element that can move (transpose) from one site on a chromosome to another.

- **transposition**
  movement of a piece of DNA around the chromosome, usually through the function of a transposable element.

- **transposon**
  a discrete piece of DNA that can insert itself into other DNA sequences within the cell. The ends of the transposon DNA are usually inverted repeats.

- **transposon mutagenesis**
  insertion of a transposon into a gene, this inactivates the host gene leading to a mutant phenotype and also confers the phenotype associated with the transposon gene.

- **transverse**
  cross-wise in position.

- **trapeziform**
  having the outline of a trapezoid, with four unequal sides, none of them parallel.

- **treads**
  the flat surfaces of stair steps.

- **treated seeds**
  seeds that have been protected against diseases.

- **tree**
  a woody, self-supporting perennial plant usually with a single main stem and generally growing more than 20 feet tall.

- **tree anchor**
  cable and peg supports fastened to a ring around a tree, to stabilise it.

- **tree guard**
  an orchard or forest nursery product of strips of various materials that circle the trunks of young trees, protecting them from rodents and mechanical damage.

- **tree line**
  the height above which trees can no longer survive on a mountainside, or the northern or southern limit beyond which only stunted forms appear on a continent toward a pole.

- **tree stakes**
  Wooden supports used to stabilise new plantings.

- **tree wrapping**
  the application of various materials to the trunks of trees to attempt to protect against damage by rabbits, children, wind, sunshine, etc.

- **trehalose**
  a crystalline disaccharide stored by many fungi instead of starch.

- **trellis**
  a frame of latticework used as a screen or as a support for climbing plants.

- **trench**
  a long cut, ditch, channel, especially man-made.

- **trench drain**
  a trench backfilled to the surface with a coarse material, such as sand, gravel, or crushed rock, for the purpose of removing excess surface water rapidly.

- **tretic**
  of conidiogenesis: The sort of blastic conidiogenesis in which each conidium is delimited by an extension of the inner wall of the conidiogenous cell. Tretoconidia are solitary or in acropetal chains.

- **triadelphous**
  with stamens fused at their filaments into three groups.

- **triandrous**
  having three stamens.

- **triangular**
  having the outline of a triangle.

- **tribe**
  a group of related genera forming a natural division within a family.

- **TCA cycle**
  Tricarboxylic Acid cycle, also known as citric acid cycle or Krebs cycle. Series of metabolic reactions by which pyruvate is oxidised completely to form carbon dioxide, and allows ATP production.

- **trichoderma**
  it is commonly found in soil, dead trees, pine needles, paper, unglazed ceramics. It often will grow on other fungi. It produces antibiotics, which are toxic to humans. It has been reported to be allergenic. It readily degrades cellulose.

- **trichogyne**
  in some algae, lichens, and fungi, a projection from the female sex organ that receives the male gamete or nuclei before fertilisation.

- **trichome**
  row of cells which have remained attached to one another following successive cell divisions. Trichomes are formed by many cyanobacteria and by species of Beggiatoa.

- **trichophyton**
  can cause ring worm, athlete's foot, skin, nail, beard and scalp infection. Reported to be allergenic. Found on soil and skin.

- **trichothecium**
  conidia dimensions 12-23 x 8-10 microns. Found in decomposing vegetation, soil, and corn seeds. The species *Trichothecium roseum* can produce a trichothecene toxin, which may be associated with disease in humans and other animals. Reported to be allergenic.

- **trichotomous**
  branching almost equally into three parts.

- **tricolpate**
  describes pollen grains with three grooves or colpi.

- **tridentate**
  the condition that exists when the large teeth of a margin have teeth, which have smaller teeth, which themselves have still smaller teeth.

- **tridynamous**
  when three stamens are longer than three others.

• **trifid**
deeply divided or left in three parts.

• **trifoliate**
having leaves in threes.

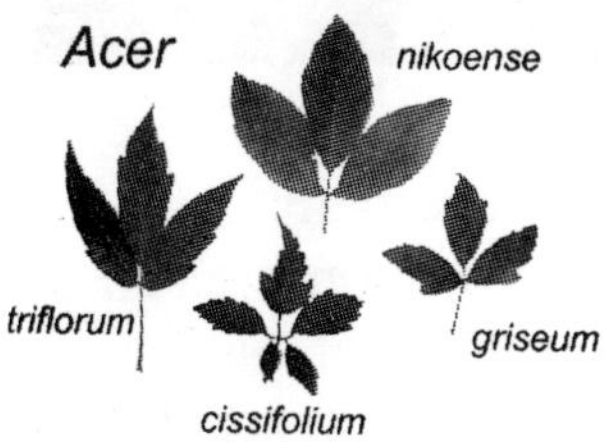

• **trifurcate**
with three branches arising from the same spot.

• **trigonal**
three-angled, triangular in cross section.

• **trilobed**
with three lobes.

• **trilocular**
with three cells.

• **trimerous**
having flower parts, such as petals, sepals, and stamens, in groups of three.

• **trimorphic**
occurring in three separate forms within the same taxon.

• **trinary combination**
trinomial, the binomial plus another term denoting subspecies, variety, forma, etc.

• **trinervate**
with three veins.

• **tripartite**
divided into three parts.

• **tripinnate**
three times pinnate.

• **tripinnatifid**
when a pinnate leaf has leaflets that are pinnate, and those are also divided into pinna.

• **tripinnatisect**
when a pinnatisect leaf has leaflets that are pinnatisect, and those are also divided into pinnatisect sections, tripinnatifid, but more deeply cut.

• **triplinerved**
describes leaves in which the midrib produces two slightly smaller side ribs near the base, the ribs forming a bird foot shape.

• **triploid**
having three sets of chromosomes.

• **triquetrous**
having three salient angles, the sides concave or channelled.

• **triradiate**
having three radiating arms or branches.

• **tristichous**
arranged in three rows or ranks.

• **trisulcate**
having three grooves.

• **tritirachium**
reported to be allergenic.

• **trophic level**
describes the residence of nutrients in various organisms along a food chain ranging from the primary nutrient assimilating autotrophs to the predatory carnivorous animals.

• **tropical**
regions of the earth lying between the Tropic of Cancer and the Tropic of Capricorn extending around the equator where the temperature and humidity are high.

• **tropical plant**
a plant that grows in tropic zones. It is what northerners make up as houseplants.

• **tropism**
the movement of an organism in response to an external source of stimulus, usually toward or away from it, including aerotropism, geotropism, haptotropism, hydrotropism, phototropism, and thigmotropism.

• **truck crops**
vegetables and fruits grown in large quantities for city markets.

• **truck garden**
a farm that grows vegetables for market.

• **true leaf**
a leaf typical of a plant that appears subsequent to the cotyledons.

• **trug**
We must include the 'English basket'. It basically is just a shallow basket for light chores, like carrying flowers and fruits and veggies.

• **trullate**
shaped like a trowel.

• **truncate**
having the end squared off or even.

• **trunk**
the major woody stem of a tree.

• **truss**
a cluster of flowers, usually at the tip of a stem or branch, generally compact.

• **tryma**
a nut with the outer shell or exocarp falling off to expose the inner nut, like the walnut, Juglans.

• **tsunami**
a great sea wave caused by underwater movement of the earth crust or volcanic eruption.

• **tube flowers**
florets with tubular corollas, usually five-lobed.

• **tuber**
a storage organ used for propagation. It may be a fleshy root (e.g. Dahlia) or a swollen underground stem.

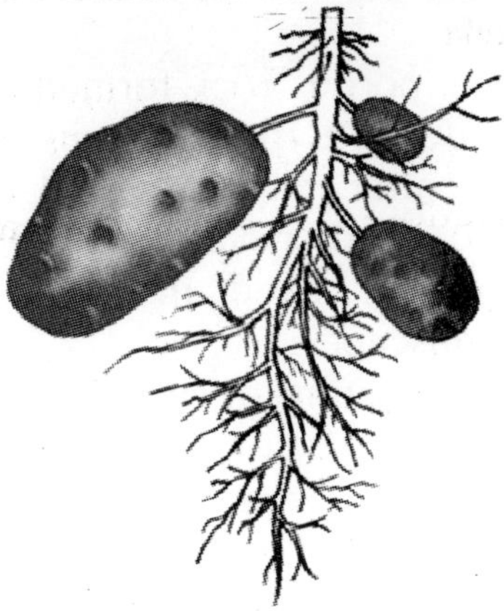

• **tubercle**
a small tuber or tuber-like (not necessarily subterranean) body.

• **tubercular**
bearing tubercles, covered with warty lumps.

• **tuberculostatic**
herbal medicine used to slow or stop the tuberculosis bacillus.

• **tuberiferous root**
a root which forms rounded knobs to store food, such as the white potato.

• **tuberoid**
a fleshy-thickened root resembling a tuber.

• **tubular**
with the petals partly united to form a tube.

- **tufa**

  1. a porous rock formed as a deposit from springs or streams, often covered in moss and/or lichen and frequently used in garden landscapes.

  2. a mixture of cement and peat moss, or other additives, formed into shapes and used as garden planters, such as tufa troughs.

- **tufted**

  having many stems in close cluster at ground level, not spreading.

- **tumefaction**

  a plant tumour or gall

- **tumid**

  turgid, appearing as if swollen from pressure within

- **tumour**

  an uncontrolled growth of tissue or tissues.

- **tundra**

  the vegetation type of very cold climates overlying permafrost and consisting of lichens, sedges, mosses, grasses, and low woody plants.

- **tunic**

  skin covering a bulb or corn, usually papery, dry, and loose.

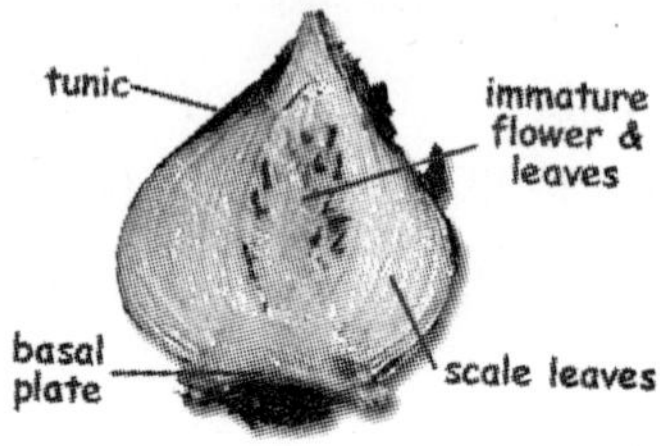

- **tunicated**

  having concentric coats, as an onion.

- **turbinate**

  top-shaped, inversely conical.

- **turf**

  a covering of mowed vegetation, usually a turf grass, growing intimately with an upper soil stratum of intermingled roots and stems.

- **turf grass colour**

  the composite visual colour of a turf grass as community perceived by the human eye.

- **turf grass culture**

  the composite cultural practices involved in growing turf grasses for purposes such as lawns, greens, sports facilities, and roadsides.

• **turf grass quality**
the composite, subjective visual assessment of the degree to which a turf conforms to an agreed standard of uniformity, density, texture, growth habit, smoothness, and colour.

• **turgid**
1. the condition of a cell, tissue or plant when it is filled with water so that it is firm, not wilted.
2. swollen, or tightly drawn, describes a membrane or covering expanded by pressure from within.

• **turgor**
the normal state of fullness and tension produced by the fluid in living cells.

• **turion**
a young, scaly, or fleshy shoot or sucker budded off from underground stems.

• **turnover**
the cycling of upper and lower strata of water in bodies of fresh water in autumn, and vice versa in spring, caused by the heating or cooling of upper surfaces. When the upper surface is cooler than the lower water, the cool water sinks.

• **twice-compound**
with the leaflets again divided into leaflets.

• **twig**
the end subdivision of a branch, a young shoot, generally applied to the growth of the past season.

• **twiner**
a climbing plant with no tendrils or suckers, in which the stem winds around other plants or objects for support.

• **twining**
clasping by winding around.

• **twisted**
the seta of many mosses twists strongly in drying. It is possible that this twisting of the seta aids in scattering the spores.

- **two-ranked**
  in two vertical rows.
- **tylosis**
  a balloon-like outgrowth from a xylem parenchyma cell that expands into and blocks the lumen of a xylem vessel or a tracheid.
- **tymovirus**
  siglum of turnip yellow mosaic virus. Member of a group of plant viruses with isometric (icosahedral) particles containing a single molecule of linear RNA, easily transmitted mechanically and transmitted in nature by beetles
- **tyndall effect**
  the scattering of light by colloids, which, for instance, causes the sky to appear blue.
- **type**
  one of several categories of herbarium specimens, chosen by taxonomists as being close or identical to that chosen by the original author.
- **ulocladium**
  isolated from dead plants and cellulose materials. Found on textiles.
- **ultimate**
  of the lowest order, the smallest division.
- **umbel**
  a part of the plant bearing flowers in which all the flower stalks are of similar length and arise from the same point.

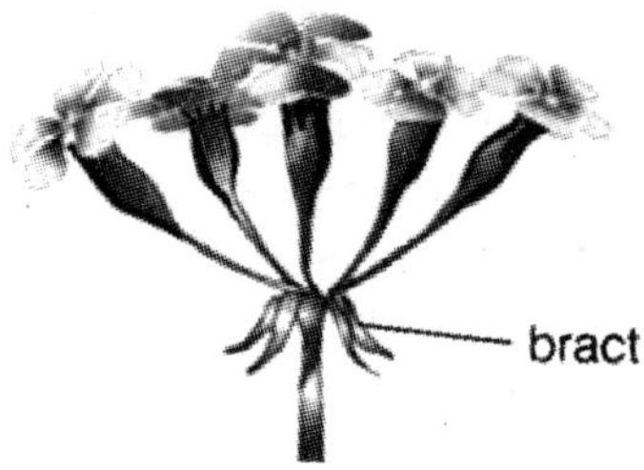

- **umbellet**
  a secondary umbel.
- **umbilicate**
  of the pileus: Having a central navel-like depression, sunken in the centre, somewhat funnel-shaped.
- **umbilicus**
  a single strand of rhizines.
- **umbo**
  a rough projection arising from the centre of an object, e.g., the thick part of the scale of a pine cone, Pinus.
- **umbonate**
  of the pileus: provided with an umbo or boss.

- **umbraculiferous**
  shaped like an umbrella.
- **unbalanced diallelic**
  the genotype involving a multiple allelic locus in autotetraploid where two alleles are represented an unequal number of times.
- **uncinate**
  hooked, curved back at point.
- **uncongeniality**
  the relationship between types of woody plants which will form a graft, but produce large swellings, severe dwarfing, or overgrowths at the plant union.
- **under glass**
  a term in older books referring to 'growing under glass', essentially growing in a greenhouse.
- **underdrainage**
  1. the placement of perforated plastic pipe or porous clay pipe underground, often with a layer of crushed stone over it, to carry away excess water.
  2. a natural or man-made system allowing excess water to be carried away from a cultivated field.
- **underplanting**
  growing short plants as a ground cover under taller plants. Under taller trees, some shrubs could be used as an underplanting.
- **undershrub**
  a low shrub that is often part of the understory growth, especially of an edge.
- **understock**
  the stock or root plant onto which a shoot has been grafted to produce a new plant in bud-grafting, especially in cases of double-worked trees.
- **understory**
  the shrubs and smaller trees between the forest canopy and the ground cover.
- **undifferentiated**
  describes parts or tissues which are very difficult to tell apart, like the tepals of tulips.
- **undulate**
  with a wavy surface.
- **unguiculate**
  contracted at the base into a claw.
- **uniauriculate**
  bearing a single auricle.

• **unicarpellate**
having only one carpel.

• **unicellular**
made of one cell.

• **unifoliate**
having a single leaf.

• **unifoliolate**
describes a leaf compound in structure but reduced to only one leaflet.

• **uniform**
always having the same shape, as opposed to dimorphous.

• **uniformity index**
the ratio between the two fibro graph span lengths (e.g., of cotton) expressed as a percentage of the longer length.

• **unijugate**
paired.

• **unilateral**
having leaves, flowers or other parts on one side only, with anthers grouped on just one side of the style.

• **unilocular**
refers to a sporangium or gametangium made of one mother-cell, which forms many spores or gametes by multiple fission.

• **uninucleate**
having one nucleus per cell.

• **union**
a graft.

• **uniovular**
with a single ovule.

• **unipolar**
at one end only.

• **uniseptate**
having one septum.

• **uniseriate**
arranged in a single row.

• **unisexual**
a flower of one sex only, either pistillate (female) or staminate (male).

• **united**
fused together.

• **unitegmic**
with a single covering over the ovule.

• **unitunicate**
having a single-layered ascus wall.

- **univalent**
  a single, unpaired chromosome.
- **unreduced**
  not having undergone meiosis, and so with the number of chromosomes per cell that is characteristic of a saprophyte.
- **upgrading**
  the reprocessing of a seed lot to remove low quality seeds and/or other materials so that the remaining seeds are higher in quality than the original.
- **upper-quartile length**
  that length which is exceeded by 25% of the fibres, by weight, in an array test specimen (e.g., of cotton).
- **urceolate**
  hollow and cylindrical or ovoid, and contracted at or below the mouth, like an urn.
- **uredinales**
  the rust fungi, obligate plant parasites in the Basidiomycotina.
- **urediniospore**
  a binucleate repeating spore borne in the uredinium.
- **uredinium**
  the sorus that bears uredinio spores in the Uredinales.
- **uronic acid**
  class of acidic compounds of the general formula HOOC(CHOH)nCHO that contain both carboxylic and aldehydic groups, are oxidation products of sugars, and occur in many polysaccharides, especially in the hemicelluloses.
- **ustilaginales**
  the smut fungi, plant parasites in the Basidiomycotina.
- **uterotonic**
  describes herbal medicines said to have an unspecified positive effect on the uterus.
- **utricle**
  a small bladdery 1-seeded fruit, any small bladder-like body.
- **vacuole**
  a cavity or vesicle in the cytoplasm of a cell usually containing fluid.
- **vadose zone**
  unsaturated zone of soil above the groundwater, extending from the bottom of

the capillary fringe all the way to the soil surface.

- **vale**
  a small valley.

- **valid**
  of taxonomic names and epithets: published in accordance with several articles of the Code of Nomenclature.

- **vallecular**
  of or near a valley or groove.

- **valvate**
  opening by valves, as a capsule, in aestivation, meeting by the edges without overlapping.

- **valve**
  in nematodes: a structure that regulates the rate and/ or direction of intake of materials (e.g., the oesophago-intestinal valve or cardia).

- **valvulated**
  having small valves.

- **vandaceous**
  from the genus name Vanda, alluding to any orchid having the characteristics of a Vanda species, large monopodial orchids such as genera Aerides, Arachnis Rhynchostylis, and Renanthera.

- **vane**
  one leaf of a windmill or pinwheel.

- **variable stocking**
  the practice of allowing a variable number of animals on a fixed area of land during the time when grazing is allowed.

- **variant**
  seeds or plants that are distinct within the variety but occur naturally in the variety, are stable and predictable with a degree of reliability comparable to other varieties of the same kind, within recognised tolerances when the variety is reproduced or reconstituted, and were originally a part of the variety as released.

- **variegated**
  plant parts having different pigments resulting in more than one distinct colour or shade on the foliage.

- **variegated leaf**
  a green leaf design which is blotched, edged or spotted

with yellow, white or cream colour.

- **variety**
  a distinct, uniform, and stable assemblage of cultivated individuals that make up a subdivision of their kind: distinct in that the variety can be differentiated by one or more identifiable morphological, physiological, cytological, chemical, or other characteristics from all other varieties of public knowledge, uniform in that variations in essential and distinctive characteristics are describable, and stable in the sense that the variety is true-breeding.

- **variety name**
  this is the scientific name or botanical name of a specific plant.

- **vas deferens**
  in nematodes: a slender, tube-like gonoduct in the male, unites posteriorly with the rectum to form the cloaca.

- **vascular**
  furnished with vessels or ducts.

- **vascular bundle**
  the conducting strand of a stem or leaf.

- **vascular system**
  the tissues in a plant that move fluids through the plant.

- **vascular tissue**
  conducting tissue with vessels or ducts.

- **vasculum**
  a metal container in which plant specimens can be carried folded in wet paper.

- **vasoconstrictor**
  an herbal agent that causes blood vessels to tighten.

- **vasodilator**
  an herbal agent that causes blood vessels to open.

- **vector**
  1. plasmid or virus used in genetic engineering to insert genes into a cell.
  2. agent, usually an insect or other animal, able to

carry pathogens from one host to another.

- **vegetation**
an all encompassing word for the plants of an area or territory.

- **vegetational analysis**
any of various methods of studying small areas of constituent plants, often counting the numbers of plants of communities, to make extrapolations to a larger area.

- **vegetative**
a cell or structure that is not producing reproductive structures, usually in the assimilative state.

- **vegetative cell**
growing or feeding form of a microbial cell, as opposed to a resting form such as a spore.

- **vegetative cover**
a soil cover of plants irrespective of species.

- **vegetative propagation**
asexual propagation using pieces of vegetation (e.g., sprigs or sod pieces).

- **vegetative stage**
the developmental stage prior to the appearance of fruiting structures. See also **pasture stage, whorl stage**.

- **veil**
the calyptra.

- **vein**
the thread of fibro-vascular tissue in a leaf or other organ, especially those which branch (as distinguished from nerves).

- **vein banding**
a symptom of virus-infected leaves in which tissues along the veins are darker green than other laminar tissue.

- **vein clearing**
a symptom of virus-infected leaves in which venial tissue is lighter green than that of healthy plants.

- **veinlet**
a subdivision or branch of a vein.

- **velamen**
the spongy, water-retaining outer layer of the aerial roots of some epiphytes, especially orchids, which are capable of absorbing atmospheric moisture.

- **velum**
  the membranous indusium in Isoetes.
- **velutinous**
  covered with fine soft hairs, even more densely than tomentose, giving the appearance of velvet.
- **venation**
  the manner in which the veins are arranged in the leaf.
- **venose**
  having veins, ribs.
- **ventilator**
  1. a device used to cool greenhouses.
  2. a perforated frame used between plant specimens in a plant press.
- **ventral**
  belonging to the anterior or inner face of an organ, the opposite of dorsal.
- **ventral surface**
  the surface of a leaf next to the stem.
- **ventricose**
  swelling unequally, or inflated on one side.
- **verge**
  border, limit, or boundary.
- **verietal**
  refers to a variety.
- **vermicast**
  the nutrient-rich by-product of earthworms, used as a soil conditioner.
- **vermicular**
  worm-shaped, thickened and bent in places. Also known as vermiculate.
- **vermiculite**
  a mineral that acts much like popcorn when heated, to form spongy, lightweight kernels, useful in loosening heavy soils and preventing compaction of the soil. Also used as a medium in which to root cuttings.
- **vermiculture**
  the raising of earthworms to aerate soil and/or produce vermicast.
- **vernal**
  appearing in spring.

- **vernalisation**
the time factor of cold days needed by certain plants to produce a bud. Apples need a certain number of cold days to create their buds.

- **verrucose**
having small, rounded processes or 'warts'.

- **versatile**
describes an anther that is attached near the middle and turning freely on its support.

- **vertical resistance**
in given cultivars: the existence of differential levels of resistance to different races of a given pathogen.

- **verticil**
a whorl.

- **verticillaster**
a cluster resembling a whorl but composed of two opposite cymes.

- **verticillate**
disposed in a whorl.

- **verticillium**
1. a fungus disease that will cause wilting and death.
2. conidia dimensions 2.3-10 x 1-2.6 microns. Found in decaying vegetation, on straw, soil and arthropods. A rare cause of corneal infections.

- **very double**
roses with a great many petals.

- **vesicate**
to blister or become blistered.

- **vesicle**
a bladder-like sac or an evanescent bubble within which zoospores mature. Any bubble-like cell or bubble-like membranous structure within a cell.

- **vesicular**
1. composed of or covered with vesicles.
2. inflated like a bladder.

- **vessel**
a xylem element or series of such elements whose function is to conduct water and mineral nutrients.

- **vestige**
a non-functioning structure that is the remnant of an organ or appendage that was once functional in previous generations or earlier stages of development.

- **vestiture**
indumentum, hair, scurf, or scales on the surface.

- **vexillum**
the standard or banner petal of a papilionaceous flower, such as a pea.

- **viability**
the possibility of germination. Seeds vary in their time of viability from a few days to 20 years, if not hundreds of years. Storage conditions will affect the viability, the best being low temperatures and humidity.

- **viable**
capable of growing or developing.

- **viable but non-culturable**
organisms that are alive but cannot be cultured on laboratory media.

- **viable count**
measurement of the concentration of live cells in a microbial population.

- **vibrio**
1. curved, rod-shaped bacterial cell.
2. bacterium of the genus Vibrio.

- **vigour**
indicative of active growth, relative absence of disease or other stresses.

- **villous**
bearing long and soft hairs.

- **vine**
a plant that trails, clings, or twines, and requires support to grow vertically.

- **virescence**
greening of tissue that is normally devoid of chlorophyll, the abnormal development of flowers in which all organs are green and partly or wholly transformed into structures like small leaves.

- **virescent**
a normally white or coloured tissue that develops chloroplasts and becomes green.

- **virgate**
wand-shaped, slender, straight, and erect.

- **virion**
virus particle, the virus nucleic acid surrounded by protein coat and in some cases other material.

- **viroid**
any of numerous kinds of small particles (250-400 nucleotides of circular, single-stranded RNA that is unencapsidated and encodes no known proteins.).

- **virulent**
capable of causing a severe disease, strongly pathogenic.

- **viruliferous**
used to describe a vector containing a virus and capable of transmitting it.

- **virus**
any of a large group of submicroscopic infective agents that typically contain a protein coat surrounding a nucleic acid core and are capable of growth only in a living cell.

- **viscid**
thick and sticky.

- **viscidium**
a sticky disc-shaped gland located at the base of the caudicle (a slender stalk-like appendage). Used to attach the pollinia (pollin packets) to an insect allowing the pollen to be carried to another flower.

- **viscous**
not free-flowing, having the consistency of syrup.

- **vitreous**
characterising slightly translucent kernels, usually, but not always, associated with hardness.

- **vitta**
an oil-tube, a structure commonly present in the pericarp of the Umbelliferae.

- **viviparous**
1. producing young plants upon various parts of the parent plant through vegetative reproduction.
2. describes seeds or fruit which sprout before they fall from the parent plant.

- **volubile**
twining.

- **volunteer**
although this could be your best friend, helping out once in a while in the garden, this is a 'not planted' specimen blown in from a neighbour's yard or a sur-

prise, from a bird or other source.

- **volunteer plants**
  plants that occur (in a population) not as a result of current seeding (of the crop under consideration) but resulting from seeds or propagative vegetative parts growing uncontrolled from previous seeding or from plants escaped from cultivation that have been scattered by natural means.

- **volva**
  a cup-like structure at the base of a basidiocarp.

- **vulnerary**
  herbal medicines that can heal wounds.

- **vulva**
  in nematodes: exterior opening of the mature female's reproductive system (female gonopore), generally appears as a transverse slit on the ventral portion of the nematode.

- **wallemia**
  conidia dimensions 2.5-3.5 microns. Found in sugary foods, salted meats, dairy products, textiles, soil, hay and fruits.

- **warm season grass**
  these grasses are grown in temperatures above 70 and 80 degrees.

- **warren hoe**
  a lightweight hoe with a triangular blade, fastened to the handle on the flat side.

- **water content**
  water contained in a material expressed as the mass of water per unit mass of oven-dry material.

- **water garden**
  any man made pool, forms, or container where aquatic plants are planted. They especially are becoming popular in the 21st Century for backyard enjoyment.

- **water sprout**
  a sucker produced on the trunk, stem, or large branch of a plant, particularly fruit trees, usually growing straight and at a right angle to the axis. These rapidly growing suckers are generally removed to allow a more open tree and to provide better nutrition to the fruit.

- **water table**
  the upper boundary of a free body of groundwater, atmospheric pressure.
- **water garden**
  planting where aquatic species predominate, or built around a pool or stream as the central feature.
- **water-retention curve**
  graph showing soil-water content as a function of increasingly negative soil water potential.
- **watershed**
  the complete area from which drains runoff to feed a stream or body of water.
- **wavy-edged**
  with shallow, rounded undulations.
- **wavy-toothed**
  wavy-edged but with more tooth-like projections.
- **wear**
  the collective injurious effects of traffic on a turf.
- **weathering**
  all physical and chemical changes produced in rock by atmospheric agents.
- **weed**
  an uninvited and usually unattractive plant that surfaces in gardens. Usually seeds are delivered by winds, but not always.
- **weeping**
  having drooping branches.
- **western blot**
  a protein blot. This term originated as lab jargon, not a generally accepted term.
- **wet beach**
  the section of shoreline repeatedly covered by water action, usually flat, firm, and without plants.
- **wet feet**
  a condition when the roots of plants are in standing water. They will eventually rot if they don't normally grow in wet conditions like aquatic plants.
- **wet weight**
  weight before drying.
- **wetable powder**
  a material that will not dissolve in water, but remains suspended in it. Most often referred to pesticides when used as sprays.

- **wetland**
  an area where saturation or repeated inundation of water is the determining factor in the nature of the soils, as well as the plants and animals living there. Included in the term are marshes, swamps, bogs, fens, bay heads, wet meadows, potholes, sloughs, bay, river flood plains, estuaries, and lake margins.

- **wettable powder**
  a finely ground pesticide or herbicide that can be mixed with water in order to spray plants. Some may be used either as a powder or dust.

- **wetting agent**
  a substance added to water that increases its solubility, allowing powders, waxes, and oils to be dissolved for spraying.

- **wheel hoe**
  a garden implement with one large wheel and two handles which is pushed to cultivate the soil.

- **whip**
  a very thin shoot with no lateral branches of a woody plant. Sometimes the first year of a grafted tree.

- **whipping pole**
  a bamboo stalk or other pole used in poling turf.

- **white rot fungus**
  fungus that attacks lignin, along with cellulose, and hemi cellulose, leading to a marked lightening of the infected wood.

- **whitefly**
  insects whose adults resemble tiny moths but are related to aphids, nymphs that suck sap and damage leaves.

- **whorl stage**
  developmental stage of a grass plant prior to the emergence of the inflorescence. Synonym (in the grass family): vegetative stage.

- **whorled**
  leaf form, where three or more leaves radiate from a single node.

- **wicking**
  conveying by capillary action, as a stem pulls moisture upward.

- **wide row gardening**
  a method in which vegetables and cutting gardens are laid out usually two to

three feet wide instead of a single file row of plants. This is efficient in spacing the plants.

- **wide-row planting**
growing the smaller vegetable crops in a space up to 3 feet across to better utilise space while reducing weeding.

- **wild type**
the phenotype characteristic of the majority of individuals of a species under natural conditions.

- **wildfire**
a fire that is out of control, such as a forest fire or a burn through a grassland, often threatening houses or farms.

- **wildflower**
plants that can be native or exotic when growing out in a non-cultivated area. They flower and are enjoyed by all. Many wildflowers of course can be cultivated in backyard gardens.

- **wilt**
a disease (or symptom) characterised by a loss of turgidity in a plant (e.g., vascular wilt).

- **wilting**
of plant disease: A symptom characterised by loss of turgor, which results in drooping of leaves, stems, and flowers

- **wilting point**
the amount of water in soil when a plant cannot obtain enough water to remain turgid.

- **wind break**
a purposeful planting of hedges and trees to protect a field, home, or garden against forceful winds, providing a shelter and preventing damage.

- **wind pollination**
the movement of pollen from the anther to the

stigma by action of natural wind currents.

- **wind setup**
wind pushing water to one side of a lake with a corresponding drop in water level of several feet on the other side.

- **window box**
a container placed under a window, very cottage-like in gardening decor.

- **windrow**
a row of cut or uprooted plants raked up or pulled together to dry and/or to facilitate harvest.

- **windward**
the side from which the wind blows.

- **windward shore**
the shore from which the prevailing winds are blowing. The boat is protected from strong wave action until the wind changes direction.

- **wing**
1. a thin, dry, leaf-like membrane found on many fruits, seeds, and leafstalks, any membranous or thin expansion bordering or surrounding an organ.
2. the lateral petal of a papilionaceous corolla.

- **winogradsky column**
glass column with an anaerobic lower zone and an aerobic upper zone, which allows growth of micro-organisms under conditions similar to those found in nutrient-rich water and sediment.

- **winter annual**
a plant from autumn-sown seed which blooms and fruits in the following spring, then dies.

- **winter kill**
a condition that happens when plants have not hardened enough to withstand sever winter conditions.

- **winter overseeding**
seeding cool-season turf grasses over warm-season turf grasses at or near their start of winter dormancy. Used in mild climates to provide green, growing turf during the winter period when the warm-season species are brown and dormant.

- **wireworms**
  long, thin, leathery, beetle larvae that attack roots and stems.

- **witches' broom**
  an abnormal form of plant growth, most common in woody plants, in which there is a profuse outgrowth of lateral buds to give a 'witches' broom' appearance. The shoots may be thickened and may bear abnormal leaves.

- **wood**
  the hard, fibrous inner tissue of the trunk and branches of a tree or shrub..

- **woodland garden**
  this garden is usually established beneath deciduous trees. It may vary from partial to deep shade and usually with plant material where roots need to remain undisturbed.

- **woody plant**
  these are usually perennial plants (i.e.. vines, shrubs, trees, and bamboos) that have permanent stems. These branches get bigger every year.

- **woolly**
  clothed with long and tortuous or matted hairs.

- **woronin body**
  spherical structure associated with the simple pore in the septa separating hyphal compartments of fungi in the phylum Ascomycota.

- **xanthophyll**
  any of several yellow accessory pigments found in plant leaves which are oxygen derivatives of carotenes.

- **xenobiotic**
  compound foreign to biological systems. Often refers to human-made compounds that are resistant or recalcitrant to biodegradation and decomposition.

- **xeric**
  of, or adapted to, an extremely dry habitat.

- **xeriscaping**
  landscaping with the use of drought-tolerant plants, to eliminate the need for supplemental watering.

- **xeromorph**
  a plant having features commonly associated with **xero**phytes, but not necessarily itself drought-tolerant.

- **xerophile**
  organism adapted to grow at low water potential, i.e., very dry habitats.

- **xerophyte**
  a plant which is able to live under very dry conditions.

- **xylem**
  tissue specialised for the transport of water and minerals upward through the plant.

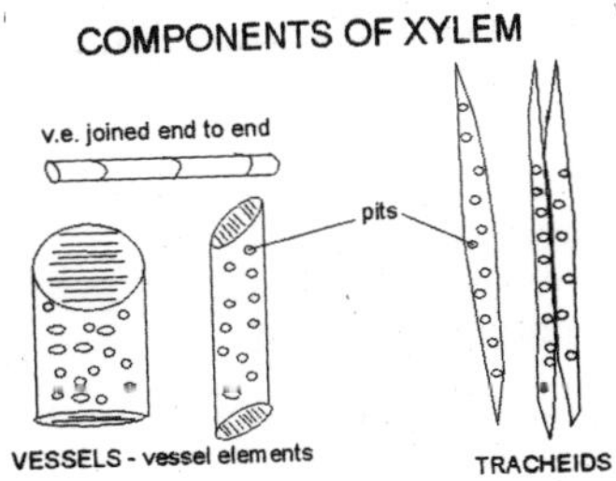

+ fibers, parenchyma, and ray cells

- **yeast**
  various yeasts are commonly identified on air samples. Some yeasts are reported to be allergenic. They may cause problems if a person has had previous exposure and developed hypersensitivity. Yeasts may be allergenic to susceptible individuals when present in sufficient concentrations.

- **yellowing**
  a symptom characterised by the turning yellow of plant tissues that were once green

- **yellows**
  any of a wide variety of plant diseases in which a major symptom is a uniform or non-uniform yellowing of leaves and/or other plant components. Yellows may be caused by fungi (e.g., celery yellows), viruses (e.g., sugar beet yellows virus), bacteria (e.g., coconut lethal yellowing), protozoa (e.g., hartrot, spiroplasmas or phytoplasmas).

- **yield**
  the aggregate of products resulting from the growth or cultivation of a crop and usually expressed in quantity per area. Commonly refers to the product or products consumed or marketed.

- **zoning**
  a ring of contrasting colour on a leaf.

• **zoosporangium**
a sporangium within which zoospores are produced. pl. zoosporangia.

• **zoospore**
an asexual spore formed by some fungi that usually can move in an aqueous environment via one or more flagella.

• **zygomorphic**
capable of being divided into two symmetrical halves only by a single longitudinal plane passing through the axis, all orchid flowers are normally zygomorphic.

• **zygogenesis**
the production of a zygote.

• **zygomycotina**
a subdivision of fungi characterised by the formation of a thick-walled resting spore formed after the fusion of two equal gametangia.

• **zygosporangium**
in the zygomycotina: a usually thick-walled, often ornamented, multinucleate resting sporangium formed following anastomosis of gametangia arising from compatible mycelia (in heterothallic species or from the same mycelium (in homothallic species).

• **zygospore**
the spore of certain fungi and algae formed by the union of two similar sexual cells. It usually serves as a resting spore and produces the sporophytic phase of the plant.

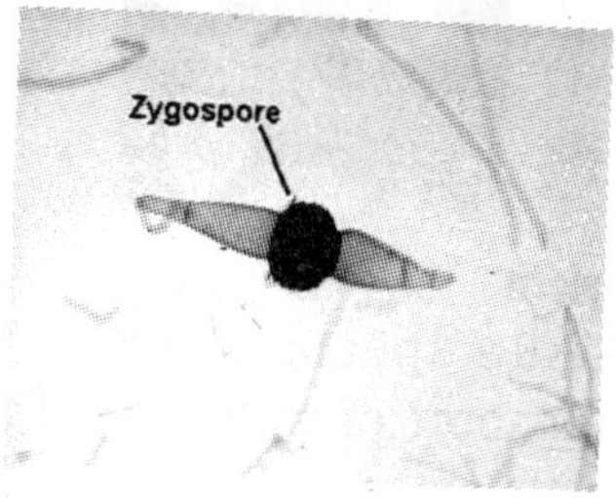

• **zygote**
a diploid cell resulting from the union of two gametes, a diploid nucleus resulting from the union of two haploid nuclei.

• **zymogenous flora**
refers to micro-organisms, often transient or alien, that respond rapidly by enzyme production and growth when simple organic substrates become available. Also called copiotrophs.